"Although evolutionary thinking in sociology predated Darwin, Alexander Riley observes that it has been only very recently that sociologists have begun to reclaim their legacy as evolutionary scientists. Providing an excellent overview of developments in evolutionary biology that began in the mid-1960s, Riley traces the recent development of what is now becoming known as 'evolutionary sociology.' Riley provides an introduction to this exciting scientific and scholarly project that will be both accessible to readers unfamiliar with the pertinent technical literature as well as stimulating and thought-provoking to sociologists who are already actively engaged in the 'second Darwinian Revolution.' Insisting that sociological analysis must be guided by the discipline of reason and evidence, Riley encourages his peer sociologists not to allow ideological and political commitments to compromise their efforts to conduct dispassionate scientific analyses. Only then can they pursue their craft in a manner that will contribute to the development of a twenty-first century social science that can fulfill the explanatory promise envisioned by its founders."

Richard Machalek, *Emeritus Professor of Sociology, University of Wyoming*

"*Toward a Biosocial Science* is an extraordinary book, entertaining, erudite, courageous, and a potential lifeline for sociology."

Edward O. Wilson, *University Professor Emeritus, Harvard University*

Toward a Biosocial Science

Sociology is in crisis. While other disciplines have taken on board the revolutionary discoveries driven by evolutionary biology and psychology, genomics and behavioral genetics, and the neurosciences, sociology has ignored these advances and embraced a biophobia that threatens to drive the discipline into marginality.

This book takes its place in a rich tradition of efforts to integrate sociological thinking into the world of the biological sciences that can be traced to the origins of the discipline, and that took on modern form beginning a generation ago in the works of thinkers such as E.O. Wilson, Richard Alexander, Joseph Lopreato, and Richard Machalek. It offers an accessible introduction to rethinking sociological science in consonance with these contemporary biological revolutions. From the standpoint of a biosociology rooted in the single most important scientific theory touching on human life, the Darwinian theory of natural selection, the book sketches an evolutionary social science that would enable us to properly attend to basic questions of human nature, human behavior, and human social organization.

Individual chapters take on such topics as: The roots and nature of human sociality; the origins of morality in human social life and an evolutionary perspective on human interests, reciprocity, and altruism; the sex difference in our species and what it contributes to an explanation of sociological facts; the nature of stratification, status, and inequality in human evolutionary history; the question of race in our species; and the contribution evolutionary theory makes to explaining the origins and the importance of culture in human societies.

Alexander Riley, Professor of Sociology at Bucknell University, USA, has read and written extensively in social theory and the history of the social sciences over the past 20 years. His work on the nature and legacy of the Durkheimian tradition is internationally recognized. Riley is the author of several books, including *Angel Patriots: The Crash of United Flight 93 and the Myth of America*, and *Godless Intellectuals? The Intellectual Pursuit of the Sacred Reinvented.*

Evolutionary Analysis in the Social Sciences
A series edited by Jonathan H. Turner and Kevin J. McCaffree

This new series is devoted to capturing the full range of scholarship and debate over how best to conduct evolutionary analyses on human behavior, interaction, and social organization. The series will range across social science disciplines and offer new cutting-edge theorizing in sociobiology, evolutionary psychology, stage-modeling, co-evolution, cliodynamics, and evolutionary biology.

Published:

Mechanistic Criminology
K. Ryan Proctor and Richard E. Niemeyer

The New Evolutionary Sociology
New and Revitalized Theoretical Approaches
Jonathan H. Turner and Richard S. Machalek

The Emergence and Evolution of Religion
Means of Natural Selection
Jonathan H. Turner, Alexandra Maryanski, Anders Klostergaard Petersen, and Armin W. Geertz

Forthcoming:

The Evolution of World-Systems
Christopher Chase-Dunn

Why Groups Come Apart
Fusion-Fission Dynamics in Human Societies
Kevin McCaffree

Maps of Microhistory
Models of the Long Run
Martin Hewson

Toward a Biosocial Science

Evolutionary Theory, Human Nature, and Social Life

Alexander Riley

Routledge
Taylor & Francis Group
NEW YORK AND LONDON

First published 2021
by Routledge
52 Vanderbilt Avenue, New York, NY 10017

and by Routledge
2 Park Square, Milton Park, Abingdon, Oxon, OX14 4RN

Routledge is an imprint of the Taylor & Francis Group, an informa business

Library of Congress Cataloging-in-Publication Data
A catalog record for this book has been requested

ISBN: 978-0-367-75098-5 (hbk)
ISBN: 978-0-367-75097-8 (pbk)
ISBN: 978-1-003-16097-7 (ebk)

Typeset in Goudy
by Apex CoVantage, LLC

To the memories of Richard Alexander, Napoleon Chagnon, Gerhard Lenski, Joseph Lopreato, and Pierre Van Den Berghe, all of whom recently left us. What we have lost cannot be replaced, even in the best of times, and we are not in the best of times.

To Ed Wilson, whose *Sociobiology*, *On Human Nature*, and *Consilience* have taught me more about the way toward a scientific understanding of humans than just about any other three books I can name.

And to the future social scientists I hope will emerge at some point, who will take the social sciences to places they cannot even see from here.

Contents

Preface
The Dismal Science of Human Nature

In a TEDx at Notre Dame talk titled "It's Not All Sex and Violence" that I have used in my introduction of sociology class for a number of years now, the anthropologist Agustin Fuentes presents a warm, cheery view of humankind and human nature. Cooperation and the ability to selflessly, peacefully get along with one another are what fundamentally mark us as a species, he argues while pacing the stage in a loose-fitting suit and shaggy, rock-star locks. Competition and conflict, and especially their more aggressive forms, are a morbid distraction from the main story of human history, which is our amazing capacity to happily work together in harmony. It is certainly an encouraging story, and my students typically agree with the charming, widely-smiling Fuentes readily and completely.

They agree with his case so easily, in fact, that they seldom note even the more glaring holes in Fuentes' argument, and the more subtle oversimplifications and misrepresentations that require more backgrounding and research to recognize escape them completely. For Fuentes' presentation is insufficiently attentive to the great amount of evidence for the parts of human nature he would like to ignore. He willfully caricatures in the most unsophisticated terms an intellectual position that is much more nuanced and consistent with evidence than his own. Of course humans do cooperate, and we do this quite well. Contrary to what Fuentes insinuates, few scholars who agree with the perspective on humankind I will present in this book make any effort at all to deny this patently undeniable part of human nature. But a full social science requires that we recognize and build into our social theory robust understanding that humans also compete with one another, and we pursue conflict, and we disagree, and we gossip, and we fight, and sometimes we even attack and kill. This is obvious to anyone paying attention, and the basic task of the social sciences is to discern what kind of intriguing creatures we are to engage in both the cooperation and the conflict and what forces and processes have made us so.

As part of his case for our peaceful nature, Fuentes presents a headline he suggests we all find easy to believe because we have been so misled by the media into a dark image of our species—"4 killed over weekend in New York City." He insists another, more impressive fact will never make headlines—"8,399,996 got along over weekend in New York City." His point is correct in one sense—extreme violence is rare in modern societies. The "why" of that fact, though, is far more complex and less flattering to Rousseauian myths about our innocent natures than Fuentes suggests. But beyond that acknowledgement, we must also note that his second headline is straightforwardly false. First, his basic math is off. It is necessary to count the killers as well as the killed among those who did not pass a peaceful weekend, so even in his framework, there are twice as many individuals as he counts engaged in aggression. But the real core of the problem with his view has to do with a much larger omitted number. Of the people living in New York City who were not among the small number of those killed and their killers that weekend, how many were involved as perpetrators or victims in thefts, robberies, extortions, kidnappings, rapes, assaults, and domestic abuse? Many hundreds, perhaps even a few thousand. How many got into fights that did not make it to the level of reporting to police? Many, many more than that. How many of those people at some point over the weekend got into an argument, or took advantage of someone else, or verbally berated someone, or contemplated or even planned some species of future conflict? Quite a large number, indeed. The reality is that competition, conflict, and aggression are much, much more common than Fuentes admits, and the reasons have to do with the fundamental forces that made us and continue to drive our behaviors and our social lives.

It's a TED talk, of course, and not a book or scholarly article, so the treatment is necessarily brisk, and students understand we grant some liberties in argument in such a medium. They nonetheless start to perk up when we go through some of the other major problems with Fuentes' case. He dismisses the Hobbesian philosophical position that a centralized political authority in modernity helps to contain what otherwise would be much more widespread crime and violence, but the evidence of what happens in modern cities when the police disappear—looting and a significant uptick in crime are the most common results—is fairly straightforwardly supportive of Hobbes. He claims aggression could not be something selected for in evolutionary terms because it is too varied a form of behavior, but in the same breath he argues that our capacity to cooperate and "be social"—something at least as varied and complex as aggression—is "wired" into us. He points to "the way we parent" as evidence of our selflessness, but he must know that the theory of natural selection has a more complicated view not only on why our parenting looks like it does, but also on what in fact motivates the behavior we call "selflessness." Parents do lovingly look after their young. And they also pick favorites among them, and sometimes abuse them, abandon them, and prevent them from being born altogether, and even kill them. The competition between parent and child begins before birth, and sexual reproduction itself bears its traces. Half of all human pregnancies fail because we have invasive placentas, and implantation of the embryo is an aggressive and invasive process by which the fetus gains increased control over its own nutrition. The tradeoff for this augmented ability of the fetus to override the mother's interest in saving her nutritive intake for herself is the higher risk of pre-eclampsia and metastatic cancer, as the invasive stem cells in the placenta are similar to metastatic cancer cells. The mother's endometrium both enables and limits the placental invasion. The embryo's efforts to increase nutrients from its mother increases her blood sugar, which might give her gestational diabetes. This is complicated stuff, and it only gets more complicated when the child emerges.

Fuentes even explicitly misrepresents a documentary on aggression in humans on which he was a scholarly consultant. Made in 2010 for the *National Geographic* channel, "Born to Rage?" is hosted by Henry Rollins and deals with the gene MAOA, colloquially known as "the warrior gene." In the effort to reject out of hand the idea—which is the subject of a serious scholarly literature—that this gene might tell us something useful about aggressive and violent proclivities in males, Fuentes declares that the only people involved with the program whose DNA tests showed the aggressive allele for the gene were several Buddhist monks and himself. A consultation of the program shows that, of the aggressive subjects presented, more than half in fact had the warrior gene, directly contrary to his claim.[1] He also fails to note that the scholarly literature indicates that the frequency of the gene varies by race, with whites less prone to have the affected allele, and nearly all the aggressive men portrayed in the program are white. This fact coupled with the accurate results from the documentary's DNA tests reveals that the individuals in the program who displayed aggressive behavior had the warrior gene at a frequency that is substantially more pronounced than in the broader white population.

This TEDx talk, though delivered by an anthropologist with training in evolutionary theory, gives a fair approximation of the worldview my students who have taken previous courses in sociology before mine have encountered in those courses. Humans, in this lens, are innately and fundamentally social and cooperative, competition, violence, aggression, hierarchy, and inequality are anomalous behaviors and forms of social organization thoroughly inconsistent with our natures, and human traits and basic aspects of social organization are mostly products of socialization and environment with relatively little contribution from crude biology. In this perspective, there is little of explanatory utility to be gained in an effort to understand our species by applying in as rigorous and unblinking a way as possible the one theory on the development of life on the planet with the most impressive track record of the past century and a half, namely, the Darwinian theory of natural selection.

This view on how to think about human nature and human social life is simply unsustainable in the long term. I present this book as a contribution to a tradition created and nourished by scholars who believe the future of social science requires us to dig much deeper than this, even if we do not always like what we find, because knowing what is the case is the best tool for successfully moving us somewhere near where we might like to go.

All writing is autobiographical, at least in some measure, and this effort at a presentation of a program for the scientific study of human nature and social life is deeply intertwined with my own intellectual trajectory over the past quarter-century. The book is my attempt to make sense of my journey through the discipline in which I was trained and toward what I have come to believe is the only hope for the future of an accurate, rigorous, and theoretically sophisticated social scientific understanding of humankind. What you will find here is basically a summary statement of my approach to the teaching of the discipline of sociology, painstakingly pieced together over the past 25 years. In that time, I have been employed by a private liberal arts institution that caters almost entirely to undergraduate students. I have therefore always seen my task in teaching sociology as requiring clear, straightforward communication, given that my students have frequently had very little previous experience in the intellectual fields from which I draw. I have brought the same attitude to this book. I have tried to avoid jargon and much of the traditional bulky scholarly apparatus of lengthy and numerous footnotes in order to try to present as clearly as possible a basic framework for the study of human behavior and human societies to an audience assumed to be interested and intelligent but that is not presumed to have a great deal of background in academic social science teaching or research. Encyclopedic works that exhaustively give the state of the literature in a given field at a given time are available; this is not one of those. There is too much material on every single major topic treated in every chapter of this book to take on the monumental task of a complete study of how each of those topics ought to be engaged with an eye toward evolutionary theory. I content myself with an introductory essay–in the spirit of others that have appeared over the years but that are now in most cases dated with respect to the existing state of knowledge—articulating the need for such a project, flying some sorties against major obstacles to such a project, and sharing a few of the most important things I have learned over the past two decades teaching and writing in sociology.

I have come to know over the years that it is a good idea not to try to inform readers as to how they should use your book, so I will avoid doing that. I would hope though that different audiences—from generally interested non-academic readers all the way to students in introductory courses in sociology or the other social sciences or even other academics who are disillusioned with mainstream academic sociology as I am—will find it of some utility. As I note at the outset of the first chapter, one of my models is E.O. Wilson in his *On Human Nature*. Wilson described that book in the following terms: "[A] speculative essay about the profound consequences that will follow as social theory at long last meets that part of the natural sciences most relevant to it" (Wilson 1978:x). As I also say in beginning the first chapter, Wilson regrettably proved too optimistic about the meeting he describes, which still has not taken place. I can however fully borrow the quoted bit for my own use if I just substitute an uncertain "if" for the "as."

Note

1 As of this writing, the documentary can still be found online here: https://www.youtube.com/watch?v=3tlq3dPmI_k&list=PLD4F9411894F6E39C.

Part 1

The Problem with Sociology and its Solution

In these first four chapters of the book, my goal is to briefly lay out the basic theoretical principles that must direct a scientific sociology. This task has both negative and positive components.

The negative component involves an examination of the failing of contemporary mainstream sociology to adopt the elements of the theoretical framework I describe here. Why is this the case? The problem has multiple aspects, to be sure, but at its root, it is a failure by sociologists to understand what a science of the human requires. The very great majority of contemporary sociological research in every mainstream methodological and theoretical tradition consists of intense focus on what are at best *proximate* cause explanations for social phenomena. This is when there is any effort to explain anything or posit causes at all—much of this work contents itself with description and explicitly eschews any effort at explanation, which is seen as an illicit or undesirable move. So, for example, sociologists who study inequality write endlessly about all the traditional sociological variables that correlate with positions in social hierarchies—e.g., levels of education, racial and ethnic identity, gender—and about the ways in which subjects understand their status and class positions and their opportunities to change their positions according to cultural narratives. But they seldom broach any questions of the contributions evolutionary science and behavioral genetics make to explaining what behavioral predilections and patterns produce social hierarchies and the inequalities they represent in the first place and what factors lead particular individuals to wind up occupying specific locations in such hierarchies. Ultimate causal explanations require a significant attempt to understand the contribution these fields make to the topic, and traditional biophobic sociology generally condemns itself to circular, irredeemably moralizing and ideological discussions of important disciplinary topics as a consequence of its theoretical and empirical myopia.

The positive component in the next four chapters involves a description of the Darwinian theory of natural selection, or evolutionary theory as I will frequently interchangeably refer to it in shorthand, a brief look at its presence in the founding generation of sociological thinkers, and an effort to show what it contributes to two fundamental questions about humankind: Why and how we are social animals, and why and how we develop moral systems to guide our sociality. I borrow freely and in some cases extensively from other scholars in presenting this framework. I am inventing nothing here, but instead am merely cobbling together in a new *bricolage* a number of the already-existing materials that I see as most productive for understanding human nature, behavior, and social life.

The position of this book is that the central reason for the decline of the discipline has to do with its refusal to come to terms with research and theoretical models of human behavior and social structure coming out of evolutionary biology and psychology, genomics and behavioral genetics, and neuroscience. Laying out the critique of the existing discipline, then, requires a presentation of the basics of what these areas of scholarship contribute to our knowledge. Such a presentation cannot but be restricted in a book of this kind, intended for a broad readership, and I limit myself to sketching out only the broadest parameters of these enormous and complex fields as they are applicable specifically to the understanding of human behaviors and social organization.

1 What's Wrong with Sociology?

In 1978, E.O. Wilson predicted, in an essay that has been an inspiration to my own thinking, the imminent merging of biology and social science. In elegant prose and rigorous logic, Wilson extended the breathtakingly ambitious effort he had made just a few years earlier to present sociobiology as a unified science of social systems in all life on Earth. In that dense 1975 book, he had dedicated but one—the final—chapter to *Homo sapiens*. But there he presented the bare bones of the much fuller case he made a few years later. The social sciences[1] would inevitably have to recognize that the insights into human nature and psychology made possible by modern biological science would have to form the basis of any systematic and scientific understanding of how humans behave and organize themselves socially. If this did not happen, Wilson suggested, sociology and the other social sciences would simply cease to have anything scientifically meaningful to say about their subject matter.

Forty years on from the publication of *On Human Nature*, there is almost no evidence to indicate that we are any closer to the moment Wilson hoped was in the offing. Let us not then beat around the bush in getting to the bad news. Sociology has lost its way, and it is now very deeply lost indeed. It is sufficiently astray that there is a significant risk that at some point soon virtually all serious people will simply stop paying attention to it, and it will only be attended to by people who in their attention reveal themselves as uninterested in scientific knowledge about human behavior and social organization.

Sociology is sometimes caricatured as "easy." It certainly is the case that, in the way it is currently taught in many institutions of higher learning, it is just that. In this incarnation, learning sociology means learning a few basic dogmatic principles. Two of the central of these principles can be summarized as follows: "Any disparity between groups is *de facto* evidence of and fully explicable by domination and injustice" and "Human beings are, because of our cognitive abilities, completely different from all other life on the planet and therefore it is not useful to apply the scientific theory most useful for understanding life—Darwinian evolutionary theory—to humans." Once you have mastered these basic dogmas, you are well on your way to being about as educated on how human societies work as any sociology major. What could be simpler? And what better explains why, in still more simplistic forms, the dogmas of mainstream sociology are so often bandied about as obvious truths in much of American media and popular culture?

Properly conceived, though, social science is hardly simple or simple-minded. In fact, in Wilson's view, the social sciences are considerably more difficult than physics and chemistry, and they therefore merit the term "hard sciences" more than those latter endeavors. What fools people into misunderstanding sociology as easy is that we all of us are human, so we have an implicit sense that we understand them—that is to say, us—well innately and without much effort. We believe we understand why we think and act, and why others do too, in relying on common sense and the ubiquitous expertise that accrues to every competent human member of a human culture just by virtue of socialization. This is, alas, a deception. It is the difficulty of cutting through that misleading sense that we know what humans are and what makes them tick to often hidden realities that creates the work for the sociologist.

At the time of Wilson's writing, sociology had only reached what he referred to as its natural history stage. Here, social science activity is entirely description of facts, with little reliable theory in which to ground them. To make matters worse, many of the facts of sociology are badly described and poorly contextualized, but the real problem is that sociological theory is so anemically weak and misguided. The building of real sociological theory will require sociology to recognize its familial relationship to related sciences. In its contemporary form, sociology has restricted itself to the study of modern human societies, claiming as the justification for this narrowing of its purview the necessary and ever-increasing professional division of labor in the contemporary university. Anthropology has a slightly larger purview, that of all human societies. But the justification for the dividing of these two sciences is intellectually wholly invalid. Human being are human beings everywhere, whether they are living in contemporary New York or in an isolated stretch of Brazilian jungle. These two sciences need to be combined, but that is only the start. *Homo sapiens* is a species of primate, and though we differ in important ways from all other existing primate species, even minimal observation reveals how much we share with them in basic behavioral predilections and social organization. And though it is still more obvious how we differ from lions, wolves, prairie dogs, dolphins, jays, and ants, enough basic principles of social life and organization unite us with them in our social natures to make it essential that sociology include knowledge about their societies in our efforts to understand our own. Sociology must also be brought together with primatology—the study of primate behavior and social organization—and sociobiology—the study of all animal social systems, as well as their biological roots—in the effort to build scientific theory.

An understanding of the proper theoretical framework within which to orient human action, Wilson argues, permits us to transform the sea of facts into a systemized body of testable propositions. In this way, a scientific sociology, while complex in the ways just noted, winds up being simpler than we might imagine, and the human animal is in at least some important ways much less complicated than some of us, in our prideful self-admiration, might prefer to be the case. What principles should a proper sociology be based on, and what should it seek to accomplish? It must start with the belief that reality can be understood objectively, although this requires considerable effort on the part of creatures like *Homo sapiens*, who are powerfully driven by emotional desires and very skilled at deceiving ourselves and others. Such effort includes not only, at the individual level, concentrated labor to be aware of biases and emotional investments in particular results and hypotheses, but also, at the collective level, structures and institutions that will allow for collective evaluation and revision of work to test, correct, and vet individual observations and studies of the human world. The goal of social science is not only description but ultimately explanation of social phenomena. So not simply: What do humans do and what do their societies look like? Instead: *Why* do humans do what they do? *Why* are societies organized in the specific ways that they are, and not in other ways? And when we see some phenomena repeatedly in many different societies, as we do, what precisely is the cause of that regularity?

Three conceptual struts hold up the edifice of a scientific sociology. We need a baseline theory rooted in systematic knowledge of human nature, that is, of the most basic principles that drive human action in the world. Darwin gave us just such a theory, and it has been developed and augmented in powerful ways in the century and a half since its initial production. Such a theory must be empirically falsifiable. Darwin's theory is, and it has been empirically tested repeatedly over time with an impressive record of passing those tests. The goal of sociological research should be to pursue reductionism wherever it is scientifically validated. We presume in advance, given our knowledge of life on the planet, that a very good deal if not all of human behavior will be reducible to the level of the psychological, or at least that much behavior will not be fully comprehensible without an understanding of basic human psychology and its origins. Finally, we seek consilience. There is nothing in the nature of human social life that leads us to believe in advance that it cannot be explained according to a basic logic that is operative in other domains of science.

This is the starting point from which a rigorous sociology ought to begin. Yet introductory texts in sociology and cultural anthropology are nearly always virtually devoid of any useful information on the evolutionary theoretical perspective on human behavior and social organization. I looked at a selection of recent such texts that are held in a student library in my own department—which is a hybrid sociology/cultural anthropology department—and found that none of them in either discipline contained more than perfunctory allusions to the study of human life as I advocate for it in this book. The sociology texts (Ritzer and Murphy 2019; Giddens, Duneier, et al. 2019) did not even have index items for "evolution," "evolutionary theory," "sex difference," or "genes." One of the sociology texts briefly mentions sociobiology, and even acknowledges that it is becoming a more common framework in some fields, but then dismisses it as inapplicable to sociology (Giddens et al. 2019:50–1). Similarly, one of the anthropology texts I examined (Nanda and Warms 2020) introduces sociobiology, in an appendix at the book's conclusion, but dismisses it flatly: "The vast majority of cultural anthropologists believe that culture is almost completely independent of biology" (Nanda and Warms 2020:383). "Cultural adaptation," this same text argues, "has freed humans from the slow process of biological evolution" (ibid.:3).

An Object Lesson from the Literature and My Classroom

Perhaps some readers believe that I unfairly caricature and criticize a sociology that does not actually exist, producing an exaggeration of what is really being done in sociology departments around the country. Allow me an illustrative example then to better make the point. In the course on sociological theory I have taught at Bucknell University for nearly two decades, I have for some years been using an article from a flagship journal in the discipline to illustrate to students just how badly sociology has gone wrong, and what dire consequences this has for the analyses it can produce of important questions about human behavior. The article in question, "Her Support, His Support" (Munsch 2015), is on the causes of marital infidelity in heterosexual relationships. The author, Christin Munsch, wields an impressive apparatus of tables chockablock with variables and their means and standard deviations, purporting therewith to bolster the claim that couples who make roughly equal incomes have the lowest rates of marital infidelity. It is not said directly in the paper, but the obvious insinuation that accompanies the analysis is an ideological one: This is the most desirable model for coupledom, which produces the most happiness for all involved.

Munsch notes in the article's introduction that infidelity seems unpopular, as most people define it negatively, and yet a fair amount of it nonetheless happens. How to explain that? She assumes that human beings of both sexes are interested in economically equitable sexual relationships. They desire for their earnings to be in approximate equity with their partner's earnings, and they are unhappy when there is an imbalance, especially when they earn less than their partner does. But it is not at all made clear why we should take this as a normative assumption of most people, even most people in advanced industrial societies like this one, and no effort is made to justify this assumption. The professional readership of this mainstream scholarly journal of sociology is unlikely to need justification of the assumption, but its absence says much about Munsch's—and mainstream sociology's—axiomatic presumptions.

It is asserted that men and women are driven for the most part in their relationships with one another by a rigid and universally applicable social gender identity, and those men and women who are in relationships in which their gender identity is most atypical will respond by engaging in behavior designed to better conform to the typical gender identity. It is then, in this analysis, the social pressure to fit comfortably into a gender stereotype that drives the dynamics of cheating. Men who feel strain with respect to their stereotypical gender identity of the breadwinner male because they are in relationships with women who earn more than they do engage in cheating to shore up their maleness. Women who feel strain in their stereotyped identity of the homemaker female because their male partners earn less than they do will be more inclined to avoid cheating precisely to better approximate the passive, sexually unaggressive female. We should therefore expect more cheating from men who make less than their partners and less cheating from women who make more than theirs. And, we are told, the data neatly show just this.

But, again, what is the evidence that men and women are motivated by what Munsch claims motivates them? It is no strength of the argument that the study omits several variables that one can safely predict would have done much to demarcate the sex difference still further along lines not traceable to socialization into gender roles. It would be interesting to see, for example, how sexual attractiveness of the subjects, evaluated simply by physical appearance, affected their propensities to cheating. Sexual selection theory—which I will discuss more in the next chapter–indicates that sexual attractiveness as evaluated by members of the opposite sex is highly likely to affect mating opportunities. Munsch also does not categorize subjects by their personal gender politics ideology, instead presuming all males and all females think uniformly about gender roles and uniformly pursue partners according to their ability to contribute materially.

Munsch inexplicably downplays an important element of the analysis that rather confounds her axiom that it is the relative relationship of male and female incomes in a couple that matters for likelihood of cheating and that the objective income has no effect on infidelity. In fact, she admits early in the article, in a one sentence statement to which she never returns or alludes again anywhere else in the course of the analysis, that this is true only for female income. As male income rises in a relationship, female infidelity goes down, regardless of the relationship of male to female income. This aspect of the reality of sex difference demonstrates with crushing clarity that the mainstream sociological theoretical model Munsch brings to bear on the data is hopelessly incomplete, and probably wrong-headed in basic ways. What does it mean that objective income level for men drives down female infidelity? It means the claim that what both sexes are fundamentally pursuing in coupledom is the same thing—income egalitarianism—is exceedingly unlikely to be true.

It is axiomatic for Munsch, and for most mainstream sociology, that men and women pursue the same goods in a long-term sexual relationship. But most people know, anecdotally in their own experience if not through an understanding of large-scale data, that men and women typically do not have precisely the same hierarchy of goods that they are pursuing in relationships. It is demonstrably true that in a small part of the globe—basically, the West and a few parts of Asia—and for a very small chunk of the time our species has been on the planet, an increasing number of women from at least a small class subset of the broader societal group pursues external employment and its rewards with as much avidity as men do. But we must not forget how exceptional this pattern is, not just historically, but even in the context of the entire global population today. In deep evolutionary time, and for reasons having to do with the sex difference itself which I examine in more detail in chapter five, our species has participated in a sexually differentiated division of labor in which men have typically worked outside the domestic sphere, for resources that they bring back to the hearth to share with wife/wives and children, while women have typically worked in the domestic realm of child care, home-making, and the gathering of sustenance from the land in the proximity of the living area. The relevant goods that males and females of our species bring to the table in sexual negotiation have historically been different, and there is little reason to believe—however much the environmental forces affecting our participation in labor have changed in recent time—that all traces of this fact of our deep history have completely disappeared.

Another telling problem emerges in Munsch's sample, which is a young one, all subjects aged 32 and younger. Older men, who are the highest earners and whose relationships typically show the most considerable spousal income gap, are simply omitted from the data. These men are also with some frequency involved with much younger spouses—often after one or more divorces—who not infrequently do not work outside the home at all. That is, the one kind of couple that arguably most starkly and bluntly illustrates historical sex difference is altogether omitted from the study. By the study's theoretical presumptions, we would expect much younger women in couples with much older men to cheat at the high end of the female scale, but she has already told us in the fleeting passage just indicated that for women a larger quantitative gap of incomes in which the man is on top lowers female infidelity.

One final—or, perhaps better, initial—point about this article and the limitations of its understanding of gender, sex, and infidelity. In the article's first paragraph, she notes a fact that one is frankly astonished to find her never return to or speculate about in any way in trying to make sense of her data or in testing her gender egalitarian theoretical framework for accurate fit with reality: As a baseline, the rate of male infidelity is in general much higher—approximately 100% or two times higher—than the rate of female infidelity. Why would this be the case, if males and females of our species are made so similarly that they pursue the same goals in sexual relationships with one another? It is almost as though something other than desire to adhere to some social norm articulated by contemporary gender egalitarian feminist sociology is driving the behavior of at least those women with high earning spouses, and perhaps of women and men as groups in their entirety. What is missing in the theoretical framework that could help us understand better what is going on here? What more do we need to know about what kind of creatures male and female humans are, what their interests are, and how they differ as a result of long-term processes that may be difficult to discern but that are absolutely essential in their makeup?

Forgotten Founders: Spencer, Pareto, and Sumner and the Early Promise

It is clear that essential intellectual foundations are missing in the Munsch article and in contemporary sociology more broadly, but this is not because they were never there in the discipline. If we look back at the founding generation of thinkers who created sociology, we find a different situation altogether. Typically, sociologists think of the founding generation as consisting of Marx, Durkheim, and Weber, and then perhaps add in a sprinkling of others as an afterthought depending on their intellectual and political skew–perhaps Tocqueville or Simmel, perhaps W.E.B. DuBois or Marianne Weber. There is however another group of thinkers from the same period, who were at mid-twentieth century seen as important founding figures in the discipline but who have disappeared over the years largely because of the biophobic drift of the discipline, who represent the discipline's early promise. Central among these figures are Herbert Spencer, Wilfredo Pareto, and William Graham Sumner. A brief look at the intellectual biographies and contributions to thought about human behavior and social organization of these three thinkers might serve us well as part of an effort to point toward a redirection of the trajectory of the history of the discipline.

William Graham Sumner (1840–1910) is today little read outside of a recent cult he has attained among libertarians for the forcible defense of a classically liberal view of labor and economics in his book *What Social Classes Owe to Each Other*. His approach to the emerging social scientific study of humankind was emphatically rooted in an appreciation for our biological and evolutionary heritage and nature. He insisted that all humans share basic needs that extend from our nature as living organisms facing challenges posed by the natural environment. Like all life, we need to discover means to extract from the world the resources required for our survival and flourishing. Modes of action Sumner called "folkways" and which he described as "like the instinctive ways of animals" emerge to meet those needs (Sumner 1960:19). These folkways, and the more elaborated morally defined categories he called "mores" that are built on top of them, are relative to a given society, but they are also all necessarily rooted in that deep human nature and therefore their relativism can be made sense of by a ranking. The ranking of human cultures then has to do with the results they yield in their capacity to provide for human needs, not with any ideological commitment. In this, Sumner's theoretical framework for understanding and comparing human societies and cultures was firmly rooted in Darwin's theory of natural selection. He was a careful student of non-Western cultures, and he recognized the excellence of many of them. All cultures can be understood as clever and inventive ways to respond to the challenges of life, but some obviously respond more effectively than others, and their members therefore flourish relative to members of other societies. Even within a given cultural and social group, there will be a range of individual sets of behavioral dispositions, and these will produce different outcomes for those individuals based on the differential consequences they yield in the eternal effort to perform the basic function of extracting the resource needed to survive from the environment.

Sumner invented the concept of ethnocentrism, which he defined purely scientifically and without a hint of the pejorative, moralizing taint it has taken on today as "the technical name for this view of things in which one's own group is the center of everything, and all others are scaled and rated with reference to it" (ibid:27) Each social group has its own norms and folkways, and so cultural diversity is an objective fact in human history. However, the natural tendency for each group is to see its own norms and culture as self-evidently superior to others, which are viewed with scorn. This is perhaps more obviously seen in primitive societies, but modern societies cannot fully overthrow this deep tendency. Patriotism is the version of ethnocentrism present in modern states: "[F]or the modern man patriotism has become one of the first of duties and one of the noblest of sentiments. It is what he owes to the state for what the state does for him, and the state is, for the modern man, a cluster of civic institutions from which he draws security and conditions of welfare" (ibid:29). Sumner was sensible to the fact that this attitude of in-group loyalty and pride could overreach into harmful extremes—chauvinism is the "truculent" extreme of this sentiment—but all groups require in-group loyalty in their members, and any that did not effectively produce it would collapse in the struggle against other groups.

If Sumner is infrequently caricatured in modern sociology thanks to the backhanded favor those who would likely dislike his ideas provide him in failing to know of his existence altogether, Herbert Spencer (1820–1903) is not nearly so lucky. Ever since his work was egregiously mischaracterized in Richard Hofstadter's *Social Darwinism in American Thought* mid-twentieth century, Spencer has been *persona non grata* in most academic social scientific circles. He shared a libertarian social philosophy with Sumner, which has contributed to the hostility he faces in contemporary overwhelmingly left-oriented academia, and his evolutionism was too unproblematically hitched to a philosophical optimism about human progress to hold up to careful historical scrutiny. But at the core of his system was a rigorous intellectual effort to construct a sociological theory that would recognize the roots all social structures must have in the basic biological aspects of the human personality. Walls made of bricks, he argued in a powerful analogy, are determined in many ways by the shape and the substance of those bricks, and so too are the patterns societies can take determined by the contours of the individual human bricks that make them up. Sociology, which desires a theory of social organization, thus must begin with an effort to thoroughly understand human nature and the behavioral and personality characteristics it imparts to the members of all empirical human societies.

Vildredo Pareto (1848–1923), roughly a contemporary of Max Weber and Emile Durkheim, entirely rejected the idealism that emerged from Durkheim's sociology as well as the insufficiently materialist critical theoretical sociology of Marx. He endeavored to put front and center in his approach to sociology the evident fact that though human behavior is frequently explained by human actors themselves as motivated by particular beliefs, most of the time when these beliefs are objectively evaluated they can be shown to be non-logical. The connection between the belief and the desired end is not apparent, and what is really motivating behavior in these situations Pareto calls residues. These are powerful predilections for action emerging from basic needs of the organism and the social animal. Among the most important residues Pareto names are the instincts to combine things, to preserve existing groups (kin, ethnic, and socioeconomic), and to express sentiments through external acts such as ceremonies and festivals. There is also a sex residue. The residues are typically hidden from the consciousness of the human individual in action. Pareto calls the beliefs that make non-logical actions appear logical to the actor, that is, the ideational self-deception that people operate in much of the time, derivations. Everyday human life is consumed with derivations, whereas, if it would be a science, sociology must concern itself with a systematic study of the residues. Such an understanding of human behavior will lead us to reevaluate much in the way of social organization. For example, Pareto famously argued that hierarchy must be understood in this light not as a deformation of proper social organization, but as a constant feature of how human beings will arrange themselves in their interactive game of the pursuit of residues.

Where Did Things Go Astray? Three Wrong Turns

Evidently, something interesting—perhaps more honestly, something devastating—must have happened between the trio of Sumner, Spencer, and Pareto, on the one hand, and Christin Munsch, on the other. We need at least a brief sketch of the history of the discipline to explain how we got to the situation illustrated by the object lesson of "Her Support, His Support." Sociology emerged in the same period of the publication of *The Origin of Species* and the rise of the Darwinian intellectual framework for studying life—that is, in the mid- to late-nineteenth century—and, as I have just described, at least some of the thinkers in its first generations recognized the promise and the need of such a theoretical framework for any scientific study of humans. Yet by the early twentieth century, three major wrong turns had cut sociology off from this rich theoretical source. Each endeavored to establish the unique intellectual status of sociology as a way of knowing, though each took its own path toward achieving that. In each case, the study of human behavior and social organization was alienated from the theoretical framework that offered the most promise for a causal, scientific sociology.

Émile Durkheim's Idealism: Wrong Turn 1

Emile Durkheim's position in the founding of the discipline of sociology is hard to overstate. Each of the four major books he published in his lifetime are still read today by undergraduate and graduate students as part of their training in the discipline, and he established the first European sociological journal, the *Année sociologique*, in 1898. Much in Durkheim's work is still of interest for the serious student of human behavior and social organization, and there is even evidence, at least in his early work, of an interest in connecting the new social science firmly to a biological core—the contemporary French scholar Dominique Guillo has written a good deal on this. But the fundamental failure of Durkheim's sociology, and of nearly all efforts to establish and build the discipline since him, even those purporting hostility to his effort, is its unrelenting idealism. Every single conceptual category in Durkheim's work—indeed, the entirety of his foundation of social metaphysics and human nature—rests on a view of human life as ultimately, in its most important aspects, fully detached from our material, biological nature. Durkheim most famously articulated this in his notion of *homo duplex*, that is, humankind as split into its less important biological side and its critically crucial mental, ideational, cultural side.

His methodological grounding of sociology as focused on what he called social facts betrays the idealism. Social facts are "way[s] of acting, thinking and feeling" defined by their externality to the individual consciousness and their coercive power to impose themselves on us regardless of our individual will (Durkheim 1982:52). Social facts are both institutions, well-established social facts that endure for some time, and what Durkheim calls "social currents," or short-lived collective enthusiasms such as sometimes move crowds to act collectively (ibid.). He insists that social facts exist as states of the *collective* mind, not merely of all the individual minds, and states of the collective mind can be observed independently of a mere addition of individual states of mind. "[L]egal and moral rules, aphorisms and popular sayings, articles of faith in which religious or political sects epitomize their beliefs, and standards of taste drawn up by literary schools" are all examples of such states of the collective mind (ibid.:55).

Just a little reflection reveals that this is metaphorical confusion. It is obviously the case that in any given modern society there is no widespread agreement about the meaning or value of many if not most moral rules, that religious and political beliefs are characterized by nuance and difference, that aesthetic tastes even within strictly defined intellectual and artistic communities are not uniform. Durkheim is inappropriately mapping a theoretical conception of how much uniformity might have been present in a hunter-gatherer society of forty people on to a kind of society it certainly cannot fit. Yet the whole methodological basis of sociology, as Durkheim defines it, rests on this confused concept. Social facts are entities with an existence independent of the human mind that have the same kind of objective influence over individuals that objects in the physical environment do. This definition calls for a rejection of methodological individualism, Durkheim argues. The biologist knows that a living entity is composed of a vast collection of inorganic "mineral substances" in interaction, yet he also understands that life is not simply a result of the total of all those inorganic substances. It is a new and original phenomenon in and of itself. So too, then, the sociologist must not presume that the entity society can be reduced to the addition of all the individual elements that make it up. But this analogy breaks down fundamentally. Both biological organisms and the chemical substances that make them up can be shown to exist. But positing in advance that society is of the same substance as an organism, an organic compound, or the elements that make up that compound is not science. Durkheim is presuming the existence of this thing that has not been demonstrated to exist outside of the collection of individual organisms that make it up. One does not have to agree with Margaret Thatcher that "there is no such thing as society" in nonetheless rejecting the assertion that society is an entity that exists apart and independent from the collection of individuals that make it up.

A still more important theoretical category in Durkheim's mature work is that of the collective representation. These, he says, "express … the way in which the group thinks of itself in its relationships with the objects which affect it," for example, a representation of a clan totem in a totemic society tells us how the group understands its origins and its place in the universe, and a prohibition indicates a collective need to preserve some aspect of the group's self-definition by indicating the kinds of things "people like us do not do" (ibid.:40–41). Collective representations, examples of which are to be found in "[m]yths, popular legends, religious conceptions of every kind, moral beliefs," are completely distinct from the individual representations that are an affair of single human minds (ibid.:41). He argues along a parallel with psychology: As the individual consciousness has individual representations, the collective consciousness has collective representations. Collective representations are a product of society, not the individual brain. Brain and mind, or material and ideal, must be neatly separated, and the latter is superior in humankind. There is, Durkheim would seem to be assuming here, a ghost in the machine, and that ghost is representation.

Durkheim wanted to define the new discipline of sociology against the established one of psychology, while, at the same time, arguing that the object of sociological research must be defined in a manner parallel to that used to define the object of psychological research. He tries to show that representations, or ideas, are not reducible to the material object that produces them, the human brain, and this brings his analysis to a level on which we have much contemporary research to use as a gauge for how close or far from the truth he was. How, he asks, can we otherwise find any ground for the memory, since the epiphenomenal theory of thought would have it that a given brain state is tied to every representation, and this would mean that any given representation must cease altogether to exist whenever the brain state is not such as to produce it? Durkheim's reasoning is universally recognized by specialists in questions of cognition as weak, especially considering what we now know, more than a century after he wrote, about brain science. He wants the reader to accept that a representation "persists after [a given condition of the brain] has passed" (Durkheim 1953:24), but how could this be so, unless one is talking about its presence in concrete symbols such as books, monuments, or other such material cultural artifacts, and here such symbols can mean nothing until some concrete mind somewhere interacts with them? A representation must be thought to exist, and this requires a state of the brain, and therefore, a brain.

Representations are ideas and beliefs that are not reducible to what is going on in individual minds. Indeed, they are not even reducible to the ideas in the heads of a collective *group* of individuals. They seem to have their residence in some ideal realm that Durkheim never effectively defines. He does however make clear that he does not even believe that *individual* ideas are reducible to the states and activities of individual brains, and he links the viability of his sociology to the irreducibility of collective representations to the associated minds of the members of a society in a parallel to the necessity he believes faces psychologists to see individual representations as irreducible to the states of individual brains. This latter is a position that is wholly untenable in light of today's cognitive science, by which we know enough about the mapping of the brain to particular realms of mental life that we can, e.g., make someone's ability to speak distort, or massively reduce his ability to feel pain in his stomach after surgery, simply by appropriately materially stimulating—through chemicals or small magnetic impulses—the correct part of the brain.

This is all of central importance to the sustainability of Durkheim's vision of sociology. His entire sociology arguably depends on his bet against reductionism of human mental activity to the science of brain structure, and we must make a fair accounting of the results of that bet in order to decide on the utility of his thinking for us today.

Is not every articulation of what Durkheim is calling a collective representation necessarily the articulation of someone, somewhere? That is, what is the entity "society" that produces collective representations if it is not simply the collection of the individual minds of the members of the social group in question? Durkheim clearly recognized the problem he faced in hypostatizing the thing "society" in the way he does, but the problem was never so much solved as simply left behind. He posited the existence of representations that are not merely individual, and argues that, if we want to properly found a sociology as psychology is properly founded, we must likewise find evidence for their existence somewhere other than in the minds of the collective group making up society. They are "partially autonomous realities with their own way of life" (ibid.:31).

Disaster, in scientific terms, awaits his argument if we do not allow the idealism he presents as self-evident. Collective representations having to do with religion and morality, he says,

> impose themselves from without upon the individual... [and] are expressly obligatory, and this obligation is the proof that these ways of acting and thinking are not the work of the individual but come from a moral power above him, that which the mystic calls God or which can be more scientifically conceived.
>
> (Durkheim 1953:25)

He intends, of course, society, in the idealist way he has just defined it, as the "scientifically conceived" substitute for God. But, it should be clear, if the argument of the relative independence of the representations from the material substrate falls, there is no such thing as society as he wants to define it; there is only the associated group of human individuals that is the material substratum from which he has just endeavored to free representations. And we are left in that case either to admit our error and become materialists, or we must plug "God" back in for "society."

How has this concept of collective representations, the linchpin concept in Durkheim's mature social theory, stood up in the light of research in the century since he articulated it? Much contemporary work in cognitive science has shown us some clear evidence that the mind is in fact reducible to the brain. We do not yet have a complete view of how that reduction works, owing to the complexity of the brain and its fabulously complicated neural networks. But everything we have learned in the last several decades indicates that we are on the right track in endeavoring to reduce the mind and all the ideas and representations it produces to the brain. Things have progressed a great deal in our research into the connections between the state of the human organismal substrate and what Durkheim calls representations. Research into various perceptual and cognitive disorders has clearly demonstrated specific brain sites as the locations at which particular ideational work is produced. Patients with damage to the upper part of the inferior parietal lobule, known as the supramarginal gyrus, a brain structure found only in humans, display an inability to perform certain skilled actions when verbally prompted to do so. The patient with ideomotor apraxia can define the term describing the act, and he has motor control of the parts of the body needed to do so, but damage to this specific part of the brain robs him of the ability to mentally represent the action in his mind prior to its physical emulation, which completely prevents him from performing the action (Ramachandran 2011:131). Such examples could be listed *ad nauseam*.

Neuroscientists working on the mapping of the neural interconnections in the brain that results from the combination of the physical structures of the brain given by basic morphology and individual genetic contribution, plus the neural pathways created by experience (known as connectomics), overwhelmingly believe that the achievement of that map, which some expect within the next century, will enable a much more complex understanding of precisely how particular parts of the brain produce certain perceptions and ideas. Perhaps the most celebrated recent example of such research is the discovery that specific neurons fire only when prompted by an image or the name of specific individuals. Though the popular press has inevitably simplified and sensationalized the research of Itzhak Fried into the "Jennifer Aniston neuron," the science behind it is mainstream among neuroscientists and suggests a tight connection between specific ideas and individual neurons or networked collections of neurons (Seung 2012:61–62). The situation today among those who study the human mind can be accurately characterized as consisting of a wide agreement that materialism in one form or another is the accurate perspective on the human species and the emergence of mind. As one prominent scholar in this field puts it, "there seems neither need, nor room, to fit any nonphysical substances or properties into our theoretical account of ourselves. We are creatures of matter. And we should learn to live with that fact" (Churchland 1989:21).

A great deal of research has also been done to test how much pre-socialized human infants have in the way of a cognitive apparatus for understanding the world. Durkheim's notion of the categories/representations as wholly or largely emerging from socialization has been tested, and the results do not reflect well on his theory. Young, presocialized infants already have mental representations of space, number, and cause, among other categories of thought (Bergesen 2004). There is even growing evidence that at least some other non-human primates (rhesus monkeys) have representations of kind and a primitive theory of mind (Phillips and Santos 2007). If this is true, a foundation of Durkheimian sociology, and much other social science, which is that human cognitive and symbolic activity is qualitatively different from anything that other animals do, simply cannot hold.

Contemporary research from cognitive science on the origins of religion advances in directions Durkheim rejected before the fact, and sheds light on the analytic power to be attained by turning to just the brain structures and the evolutionary pathways by which they were constructed that Durkheim thought sociology had to reject. Scott Atran (2002) and others argue that there is reason to believe that a selective advantage for a "hair-triggered agent detector," that is, a very high sensitivity to the presence of other sentient agents in one's vicinity, is likely one of the core evolutionary elements that made religion possible. In the late Pleistocene age, when *Homo sapiens* was emerging, it is quite likely that the most dangerous, deceptive predator of our species was other members of our species, and it would have been an advantage to detect agents even in some cases of uncertainty and foreboding where there in fact were none ("better safe than sorry"), and a disadvantage to those individuals who were less capable of detecting agents. Supernatural entities are not agents, insofar as they do not exist, but a capacity for belief in such entities may have been a by-product of this heightened sensitivity to real agents. So, far from the proposition of Durkheim that religious beliefs start as a collective enterprise, it may well have been that such beliefs were a consequence of other, more primary cognitive skills that were directly related to organismal adaptation and competitive fitness.

Atran also hypothesizes that what we have learned about memory and communicative transmission of narratives from evolutionary psychology seems to indicate that beliefs about e.g., invisible creatures who transform themselves at will, i.e., beliefs that shake our basic notions of the nature of reality, are more easily remembered and passed along as cultural memes than more random departures from common sense. Such beliefs therefore would have been relatively successful at reproducing themselves. We know from research in developmental psychology that even young children have universal expectations about the nature of the world that include basic, intuitive categories such as person, animal, and plant. It is shocking, and attention-arresting, even to children, when we are presented with systematically expounded ideas that are counterintuitive to this ontology. Such ideas are easily learned and transmitted, in fact, even more easily learned than intuitively consistent items in narratives. For example, when Native American folk tales were told to participants and the latter were then asked to remember them, they recalled over 90% of the minimally counterintuitive (i.e., supernatural) claims, but only about 70% of the intuitive ones (Barrett and Nyhoff 2001). A view of the mode of functioning of the individual human mind from the perspective of contemporary neuroscience gives us powerful tools that challenge Durkheim's framework.

The problems presented by Durkheim's wrong turn for modern mainstream sociology are multiple and catastrophic. In his work, the entity "society" becomes a purely ideal object. It exists in the form of representations, or ideas, but these are ideas contained not in the heads, that is, brains, of individual human beings, but in some ether that is poorly defined and defended. Culture is also for Durkheim a purely ideal entity, with no connection to the material world other than in its presence in the printed, engraved, and other made objects that bear cultural messages and symbols. It is most definitely not something that we should seek in deep behavioral needs of humans and the evolutionary processes that put them there. Finally, Durkheim sees human interests as fundamentally about cooperation and the expanding well-being of all. Though he wrote famously on the origins of religion in tribal solidarity, he also wrote on the inevitability and indeed the moral superiority of a globalist understanding of human interests, in which tribal connections would be broken down and all would come to see that their interests were incompatible with the competition, exclusivity, and incommensurability of interests in the natural world.

Again, to give Durkheim his due, he certainly showed some interest over the course of his intellectual career in evolutionary framings. His book on religion posits an essentially evolutionary model for understanding the development of religion over time. But it misses an opportunity by failing to discuss the evolutionary forces that almost certainly led to religion in the first place. The contradiction at the heart of mainstream sociology is pristinely clear in Durkheim. "Sociology's enigma," as Charles Lemert has described it, is the desire to have a science separate from, independent of psychology—and, it goes without saying, biology—and doing that requires the ideal things—society, culture, social structures—that one finds everywhere in the early pages of the textbooks, yet everyone who is honest knows those things cannot be shown to exist empirically apart from the concrete behavioral patterns of individual human beings. These ideal objects must be inferred, invented, from particular ways of spinning empirical facts (Lemert 2011:79). Sociology could have avoided all this had it only started with the material fact that is the essential starting point of any study of human beings, the object Durkheim relegated to the unimportant half of *homo duplex*: The evolved biological organism that is the individual animal of the species *Homo sapiens*.

Boasian Cultural Anthropology and Marxian Critical Theory: Wrong Turns 2 and 3

As sociology began propitiously with thinkers like Spencer and Sumner before turning later to idealism, so anthropology took a wrong turn in the same direction after its founders had established the discipline in the firm soil of evolutionary theory. The Englishman E.B. Tylor and the American Lewis H. Morgan each developed a framework for studying culture as an evolutionary phenomenon, though both were too caught up in the widespread nineteenth-century mistake of thinking of evolution as progress rather than simply as adaptation through variation within varying ecologies. But the thrust of their work was negated in the early twentieth century by a revolutionary and anti-sociobiological thinker who moved the discipline in a different and fundamentally incompatible direction.

Franz Boas (1858–1942) was, from an early age, drawn to questions of what today would be labelled "social justice." He was an early champion of the notion, not fully birthed until the late 1960s, that African history was as great and glorious as the histories of Asia or the West. His views on this topic, presented at the invitation of W.E.B. DuBois in 1905 to a black audience at Atlanta University, astonished his American host with their radical thrust (Degler 1991:76). Boas championed the independent explanatory power of culture in the effort to understand human behavior and outcomes. He rejected altogether the then dominant notion in anthropological science for explaining distinctions in human populations, which was biological race, and aimed to reorient anthropology on a culturalist base. His students Ruth Benedict (1887–1948) and Margaret Mead (1901–1978) pushed the idealist positions of their teacher to still greater extremes. Benedict's *Patterns of Culture*, published in 1934, forcefully rejects not only any real biological contribution to human social life, but also the idea that cultures can be usefully understood in evolutionary relief. The cultural relativism she introduced is not simply a scientific tool to prevent intellectual bias, but also a political one to fight the "nationalism and racial snobbery" that she claimed had emerged in response to the fact that "modern existence has thrown many civilizations into close contact" (ibid.:206). A year later, Mead's book *Sex and Temperament in Three Primitive Societies* shook up middle-class American readers still further by pouring boiling acid on one of the most basic aspects of biologically-based human nature, the sex difference. In this study of three New Guinean societies, Mead claimed to have found stupendous variation on sex roles and aggression that proved innate sex characteristics a non-factor in the face of the superior force of culture. In one of the societies she described, both men and women showed low levels of aggression; in another, both sexes were extremely aggressive; in the third, women were more aggressive than men. Culture, in other words, has the final say, and it can say whatever it wants, purported biological realities notwithstanding. Human nature for Mead was "almost unbelievably malleable" (ibid:134). A subtext of her book's argument is the idea that Western notions of the sex difference are not to be considered any more accurate or valuable than the notions of gender roles among the Arapesh, the Tschambuli, or the Mundugumor.

Under the spell of this intellectual worldview, by mid-century it had become *de rigueur* in anthropology to assert, as Leslie White and Ashley Montague did, that there was no human nature and that culture was the only thing to be considered in considering human behavior. "[E]verything [man] is and has become he has learned, acquired, from his culture," wrote Montagu (ibid:209). The two basic ideas that emerged from the development of twentieth-century American anthropology were cultural plasticity and cultural relativism. Human cultures are infinitely varied, and they are equally viable and valuable. It is no small matter that neither of these claims is true, as I will discuss in this book's final chapter.

From the cultural anthropological tradition, we get a view of social science as a merely interpretivist endeavor. The most recent developments in this tradition willingly descend into a more or less consistent subjectivism. Reality can be understood only in the form of interpretations of meaning, these are ultimately tied to subjective positions, no two observers share the same perspective, and so there is no real possibility of objective knowledge of human life. Indeed, any confused effort to pursue this will only end in distortion. All is interpretation, and the goal of sociology is not the explanation of social phenomena but only its description, with no hope of completeness and ultimately only the goal of Clifford Geertz's "thick description," which is aesthetic and not scientific. Social science is, in this vision, the natural ally not of biology but of disciplines in the humanities such as philosophy, religion, and literature. A destination more distant than this from a full-blooded scientific sociology can scarcely be imagined.

A final wrong turn begins with Karl Marx (1818–1883), who famously claimed intellectual indebtedness to the work of Darwin in a number of remarks about Darwin's ideas. Marx even imagined he was bringing to social science the same intellectual perspective Darwin had brought to biological science. It is true that Marx developed a materialist understanding of human action and interests that was in some ways closer to evolutionary thinking than the idealist theorizing done by many in the founding generation of sociologists. However, it is also quite clear that Marx rejected or was ignorant of basic elements of the evolutionary theory proposed by Darwin. Instead of individual human organisms as the objects under selection, Marx saw groups as the entities that pursued interests, struggled against one another, and were shaped by the evolutionary process. Those groups were not even actual societies that lived together, but abstract groups he called social classes that were distributed across the globe and historical time. Marx paid no attention whatever to most of the key features of Darwin's work, e.g., the sex difference and sexual selection and the way existence under different ecological pressures would produce differentiation in separated regional human populations are wholly absent in his work.

It is the moralizing side of Marx's project that derails it utterly from any trajectory that might lead to scientific knowledge of the operation of human society. If the failure to attend to basic elements of Darwinian evolutionism handicapped Marx's ability to deeply explain historical human action, his attachment of the analytic elements of his thought to an incoherent utopian political dream ensured that Marxian social science—as well as all the later variants descended from it in the way of various critical sociologies—would fall well short of the rhetoric of its founder. For the first half century of sociology's institutionalization, Marx's thought was at the periphery of the theoretical mainstream, which was dominated by various functionalisms. Though Marxian and Marxian-derived theoretical frameworks have in the past fifty years attained a position of some dominance in American and many European sociologies, they have not overcome the limitations of the founder.

Indeed, critical sociology after Marx has rejected most of the more viable elements of Marx's work, keeping only the empirically incoherent idea of human history as the struggle of transhistorical groups of oppressed and oppressors and the moralizing commitment to see sociology as more a form of politics than a scientific endeavor to understand human behavior and social organization. Critical sociology argues that our investigation of the human world is inevitably driven by moral and political principles. Here too, as in the interpretivist strand that emerged from Boas, and in opposition to the rhetoric of Marx if not to the actual tenor of his work, the effort toward objectivity is rejected, and it is seen as little more than a masked effort at domination. Sociology is here first and foremost a contribution to political and social change. The notion of science was invoked frequently in the Marxian tradition, but with precious little evidence of any real investment in scientific reasoning or method. Since Marx, critical sociology has frequently been unapologetic in sternly rejecting, and frequently even denouncing, the idea of a scientific sociology as just another mechanism by which sexist, racist, and classist dominant powers seek to maintain their privilege and authority. A peculiar contradiction of a fair amount of contemporary critical sociology is the rhetorical effort to claim the mantle of science—always defined in ways wholly antithetical to science as practiced in the accomplished scientific disciplines—in order to benefit from the status of such a claim, while simultaneously invoking relativist and postmodernist criticisms of the epistemological uniqueness of science. Nothing demonstrates the poverty of contemporary sociology more starkly than this spectacle.

On Human Nature, Again

I do not want to leave the impression that I unduly caricature sociology and its ignorance, even hostility with respect to evolutionary thinking. There are serious thinkers in contemporary sociology who have spent time with evolutionary and sociobiological approaches to studying humans and nonetheless taken up positions different from my own. Jeremy Freese, for example, has written a good deal on the question of whether and how much sociology ought to incorporate evolutionary thinking. His perspective on the sociobiological view is generally critical, and he has gone to the trouble to attempt to empirically test some elements of human behavior to see if evolutionary frameworks are helpful in understanding them. He and a colleague purport in one such work (Freese and Powell 1999), for example, to refute the Trivers-Willard hypothesis by showing that there is no evidence parental strategies for favoring male or female offspring are affected by the relative status of parents. But they test it on a sample of contemporary Americans, and they do not acknowledge that all serious evolutionary analysis recognizes that culture mediates the evolutionary forces. Indeed, the most rigorous such frameworks make clear that human beings have been living for at least the past ten thousand years in environments that are radically different than the natural environment in which we spent most of our time as a species and in which our evolutionary predecessors also lived. Behavior can only be determined to be adaptive when we consider it with respect to this natural environment, which is the sub-Saharan African savanna during the Pleistocene era stretching from about two and half million years ago until around 10,000 BCE. Thus, determining the adaptiveness of human behaviors today can be difficult, and determining ultimate causation of many behaviors may be made next to impossible. In this and his other writing on the topic Freese, ultimately like nearly all sociologists, turns out to be insufficiently motivated to give evolutionary sociology the fair intellectual examination it merits.

And most of the rest of the discipline goes nowhere near the level of engagement with this body of intellectual tools evinced by Freese. Contemporary sociology is, as the Munsch article demonstrates, filled with inquiries into questions about human nature that it is patently unequipped to handle competently. It does not have the excuse that critical perspectives on its endeavors such as the one represented by this book have been missing. The position I sketch here is hardly novel. Indeed, much of what I will say here has quite literally been said by other sociologists and anthropologists who have come before me over the last several decades to alert sociologists to their wrong turn, to no avail. These predecessors include Joseph Lopreato and Timothy Crippen, who some twenty years ago wrote the book that I have used for more than a decade as my own textbook in my introductory sociology course, Gerhard Lenski, whose explicitly ecological-evolutionary theoretical model is one of the rare systematic efforts in modern sociology to think human nature *à la longue durée*, Lionel Tiger, Robin Fox, Donald Symons, Pierre Van Den Berghe, Richard Machalek, Jonathan Turner, and Alexandra Maryanski, to name just some of the most prominent. That their warnings to the discipline have by and large not been heeded bodes ill for my own minor intervention. Yet in some other corners of the social sciences—evolutionary psychology, especially, but behavioral economics is another vibrant example—this perspective thrives, and it is likely that other spaces in the social sciences will open up to it as the evidential base grows, which it will certainly do.

Note

1 I will follow the example of the first generation of sociologists in using the two terms "social science/s" and "sociology" interchangeably. Of course, I recognize the significant theoretical and other differences between the contemporary disciplines of sociology and e.g., economics and political science, and I am quite aware that in some social science fields evolutionary theory has been much more effectively integrated into routine research than is the case in sociology, but the general point I make in this book about the utility of the unified theory I discuss applies across the board.

References

Atran, Scott. *In Gods We Trust: The Evolutionary Landscape of Religion*. New York: Oxford University Press, 2002.

Barrett and Nyhoff. "Spreading Nonnatural Concepts." *Journal of Cognition and Culture*, 1 (2001):69–100.

Benedict, Ruth. *Patterns of Culture*. London: Routledge and Kegan Paul, 1935.

Bergesen, A.J. "Durkheim's Theory of Mental Categories: A Review of the Evidence." *Annual Review of Sociology*, 30 (2004):395–408.

Churchland, Patricia. *Neurophilosophy: Toward a Unified Science of the Mind-Brain*. Cambridge: MIT Press, 1989.

Degler, Carl. *In Search of Human Nature: The Decline and Revival of Darwinism in American Social Thought*. New York: Oxford University Press, 1991.

Durkheim, Émile. *The Rules of Sociological Method and Selected Texts on Sociology and its Method*. Edited by Steven Lukes. Translated by W. D. Halls. New York: Free Press, 1982.

—. "Individual and Collective Representations." In *Sociology and Philosophy*. New York: Free Press, 1953.

Freese, Jeremy and Brian Powell. "Sociobiology, Status, and Parental Investment in Sons and Daughters: Testing the Trivers-Willard Hypothesis." *American Journal of Sociology*, 106 (1999): 1704–1743.

Giddens, Anthony, Mitchell Duneier, Richard Appelbaum, and Deborah Carr. *Essentials of Sociology*, 7th edn. New York: W.W. Norton & Company, 2019.

Guillo, Dominique. "La Place de la biologie dans les premiers textes de Durkheim: un paradigm oublié?" *Revue française de sociologie*, 47, 3 (2006):507–535.

Hofstadter, Richard. *Social Darwinism in American Thought*. Philadelphia: University of Pennsylvania Press, 1944.

Lemert, Charles. *The Structural Lie: Small Clues to Global Things*. New York: Routledge, 2011.

Lenski, Gerhard. *Ecological-evolutionary Theory: Principles and Applications*. New York: Routledge, 2005.

Lopreato, Joseph and Timothy Crippen. *Crisis in Sociology: The Need for Darwin*. New Brunswick: Transaction, 1999.

Mead, Margaret. *Sex and Temperament in Three Primitive Societies*. London: Routledge and Kegan Paul, 1935.

Munsch, Christin L. "Her Support, His Support: Money, Masculinity, and Marital Infidelity." *American Sociological Review*, 80, 3 (2015):469–495.

Nanda, Serena and Richard Warms. *Cultural Anthropology*, 12th edn. New York: Sage, 2020.

Pareto, Wilfredo. *Selections from his Treatise*. Edited and introduced by Joseph Lopreato. New York: Thomas Y. Cromwell Company, 1965.

Phillips, Webb and Laurie Santos. "Evidence for Kind Representations in the Absence of Language: Experiments with Rhesus Monkeys (Macaca Mulatta)." *Cognition*, 102, 3 (2007):455–463.

Ramachandran, V.S. *The Tell-Tale Brain: A Neuroscientist's Quest for What Makes Us Human*. New York: W.W. Norton & Company, 2011.

Ritzer, George and Wendy Wiedenhoft Murphy. *Introduction to Sociology*, 5th edn. New York: Sage, 2019.

Seung, Sebastian. *Connectome: How the Brain's Wiring Makes Us Who We Are*. New York: Mariner Books, 2012.

Spencer, Herbert. *The Study of Sociology*. London: Henry S. King & Company, 1873.
Sumner, William Graham. *What Social Classes Owe to Each Other*. New Haven: Yale University Press, 1925.
—. *Folkways*. New York: Mentor Books, [1906]1960.
Wilson, E.O. *Consilience: The Unity of Knowledge*. New York: Vintage, 1998.
—. *On Human Nature*. Cambridge: Harvard University Press, 1979.
—. *Sociobiology*. Cambridge: Harvard University Press, 1975.

2 The Basics for an Evolutionary Sociology

Theodosius Dobzhansky, one of the most venerated figures in modern evolutionary biology, famously wrote that nothing in biology makes sense except in the light of evolution. The same thing is true in the social sciences, though most contemporary social scientists give little evidence of recognizing it. Evolutionary logic is the fundamental organizing theoretical framework for the study of humankind, and the mechanisms by which humans evolve must be part of the study of human nature and social life. E.O. Wilson has recently summarized this view in his usual crisp and eminently quotable way:

> Every part of the human body and mind has a physical base obedient to the laws of physics and chemistry. And all of it, so far as we can tell by continuing scientific examination, originated through evolution by natural selection.
>
> (Wilson 2019:17)

How is it possible, then, that virtually the entirety of the contemporary discipline of sociology rejects this notion, or at least displays an ignorance of and disinterest in evolutionary theory that make it impossible to understand its utility in the social sciences? A more personal version of the question that I have posed to myself many times is: How did I fail to clearly understand this as a young student of sociology, despite a knowledge of evolutionary theory, and instead spend more than a decade working outside this framework, without the slightest effort to seek integration of what I knew about the most basic theory for understanding the emergence and development of all life on earth and what I researched, wrote, and taught as a sociologist?

The answer has largely to do with the development of sociology as a discipline and a profession. Undergraduates can easily earn bachelor's degrees in sociology without ever encountering evolutionary theory or sociobiology. Graduate students in sociology can also make their way entirely through a Ph D in the discipline without ever having to intellectually confront this paradigm. One of the things I have noticed in the years I have spent teaching material related to the content of this book is that many—not nearly all of them, but quite a few—students have a powerful allergy to this information. They constantly and confidently invoke the categories of mainstream sociology in denigrating evolutionary theory, even when they are not sociology majors, because these categories are now the grist of the public mill of knowledge about such things. You name the question, they have ready answers to it, even if they have never read any serious scholarship on it. Why are there few women in physically risky jobs? Because of sexism. Why are there consistent in-group/out-group sentiments in multicultural societies like this one? Because of racism. Why are men and women typically so different from one another, not just in their physical attributes, but in their behaviors, their interests, and their personalities? Because of anti-feminist gender socialization. Why are the overwhelming majority of human beings on the planet sexually interested in members of the opposite sex, and why do the overwhelming majority identify as a member of the gender that corresponds to their sex? Because of homophobia and transphobia. Why do most people demonstrate antipathy for hierarchy only when they are in a subordinate and not in a superordinate position in the relevant hierarchy? Because they have been brainwashed by capitalism. These answers are, for many at least, much more comforting—and they are certainly more easily reached, with significantly less reading and study—than those suggested by evolutionary theory.

I read Darwin and some modern evolutionary biology as an undergraduate and found it fascinating, but I was never presented with any substantive opportunities in my study of the social sciences to see its viability for that application. In graduate school, I was in the same institution as some great figures in contemporary neurophilosophy, Patricia and Paul Churchland, and I read them then, but the widespread belief among my sociology colleagues was that there was simply no real way to connect what "we" did with what "they" did. By the merest accident, I fell upon copies of E.O. Wilson's *Sociobiology* and *On Human Nature* in one of the used bookstores in San Diego that I haunted in those days and devoured both at the same time I was finishing my dissertation on the history of Durkheimian social thought. But the two worlds seemed entirely distinct to me then. And I never encountered, in my decade in higher education as an undergraduate and then graduate student, a single course—not even a single text or mention in any course—on the utility of evolutionary thinking in the study of human beings. I was introduced to Darwin and more modern evolutionary biologists such as Stephen Jay Gould and Richard Lewontin—and it is telling that it was these two specifically I was given to read, as we will see shortly—in undergraduate courses in biology, but there were no efforts made in those courses to connect that material to the study of human behavior or social activity. In sociology, I read the canonical material—Marx, Weber, Durkheim—and then moved on to contemporary treatments of sociological questions that were as far away as imaginable from evolutionary theory. This is all to say that the discipline of sociology is a socializing, professional organizational set of structures that makes it quite difficult in the vast majority of cases to ever see what needs to be seen. But this is only the first part of a big problem.

The Logic of Evolution *via* Natural Selection in General and in Humans

Americans do not know very well at all the most basic and singularly productive scientific theory involved in the explanation of human origins and human nature. Survey data suggest that Americans are woefully underinformed about how evolution works on living organisms, and it is not just Americans without college education who are benighted in this respect. While at the level of surveys, it is certainly indicated that the more educated among us show somewhat greater awareness of most basic elements of the theory of natural selection, even educated Americans typically know little about how the theory has been applied to understanding our species and our social lives. Those educated Americans are also quite frequently at least as hostile to Darwinian and sociobiological frameworks applied to humans as their less-educated fellows, if sometimes for different reasons. Whether on the political right or left, religious or irreligious, we frequently reject the very idea that we might be driven in some fundamental ways by material, historical processes that elude our direct consciousness in action and that undermine some of the most long-lived explanations—and our cherished hopes—for our behavior and our character.

There are several cognitive biases that tend to prevent individuals from understanding the logic of natural selection and make them skeptical about its explanatory power (Legare et al. 2018). These include psychological essentialism (the belief that some common, stable, and immutable essence is what defines members of a species) and a profound tendency in humans toward teleological thinking (the belief that things exist because they serve some purpose, rather than as a result of system or chance processes). Among social scientists, there are further obstacles to understanding natural selection in a human explanatory context, including most fundamentally the belief that culture and social learning are mechanisms that elude natural selection altogether. Most educated adults, and probably most professional sociologists, understand that all life on the planet has a shared ancestry, and any two organisms should be understood as having a shared ancestor at some point back in the past of the tree of life, however distant. But the utility of this logic in the study of humans is overlooked, or even rejected outright, by most contemporary sociologists. My own anecdotal experience over two decades as a professional sociologist indicates that the typical member of my discipline knows startlingly little about the theory of natural selection, human evolutionary history, or evolutionary theoretical efforts to explain human behavior and social organization.

But why should we not accept the inclusion of our species in the explanatory logic with the best record to date of explaining how living things operate to the fullest degree? Why not recognize what we share with the rest of life on earth, which is not only our rooting in the material, biological processes that drive our brains but also the logic of natural selection that built them? I sympathize completely with Patricia Churchland's reaction to such knowledge: It brings me a certain joy to recognize my deep relationship to the rest of life on earth (Churchland 2013:20).

Some critics of the approach taken in this book claim that all the relevant work natural selection did on humans was done long ago, and that in any event the tens of thousands of years since humans began agricultural activity is not a sufficient period of time to have produced any modifications generated by selection. This is not so. There is ample evidence that our species has been and still is operating under processes of natural selection. The widespread emergence of lactose tolerance, which is produced by a single gene mutation now characteristic of 30–35% of the global human population, happened in just the last 10,000 years. The mutation spread in populations that domesticated cattle for labor and food because, in conditions of significant food scarcity and deprivation, the people in those populations with the mutation of the gene that allowed them to ingest dairy products into adulthood had a caloric advantage over those who could not, and those individuals with the mutation reproduced themselves more successfully and eventually became majorities in their regions of the world. There is also evidence that genes implicated in impulsive behavior and novelty-seeking are under positive selection at present, and this might be connected to, among other phenomena, the massive rise in the diagnosis of ADHD in the young (Esteller-Cucala, Maceda, Børglum et al. 2020). There are many other examples.

That an evolutionary framework is useful for understanding profoundly important human problems is not hard to demonstrate. The field of medicine now accepts as well-established individual and population differences at the genetic level that are of the utmost importance in treating patients. There is, for example, a gene that codes for the chemokine receptor that the human immunodeficiency virus, or HIV, uses to enter host cells. One allele of the gene prevents HIV from entering cells. This allele is modestly present (around 9%) in European populations, but completely absent in Asian and African populations. This is evidently of importance in the medical work of understanding the spread of infections in different populations. At least some sociologists would likely accept this as a reasonable intrusion of evolutionary logic into the study of an aspect of human life. But, they might quickly add, this example is as much to do with the evolution of a virus as that of humans. Given what (they think) we know about what is selected by nature, why would we want information on the evolutionary origins of human behavioral predilections? Perhaps, this critic might well argue, such information, if available, would be interesting from a merely explanatory framework, but would it not then close off the reasonableness of asking how to alter those behaviors? They are after all "natural" behaviors. Is an evolutionary perspective on human nature and action just at bottom a way of excusing whatever humans are and do as natural and therefore good, or at least as normal?

This is the naturalistic fallacy: Whatever is determined to be the normal, natural condition must be tolerated, even acknowledged as the desirable, best outcome. But this claim clearly does not hold up to scrutiny. An understanding that human beings are fairly well-suited in terms of our evolutionary psychology for interpersonal conflict up to and including lethal violence does not mean we cannot prosecute and punish those who assault and kill others. And it certainly does not mean we cannot endeavor to produce conditions in which humans will do this to one another less frequently than they currently do. Understanding that e.g., murder ideation is as common as research demonstrates it is (Buss 2006) helps us to be aware of the real magnitude of the problem, and to better prepare ourselves in interactions with others. It may also help us to understand what kinds of interventions are more likely to have the desired effect on the predilection to the behavior. As Thornhill and Palmer argue in their study of rape,

> individuals who really understood the evolutionary bases of their actions might be better able to avoid behaving in an 'adaptive' fashion that is damaging to others…knowledge of the self as having evolved by Darwinian selection provides an individual with tremendous potential for free will.
>
> (Thornhill and Palmer 2000:154)

In the exact same way that we better prepare medical responses to HIV when we understand what processes produce its ability to mutate, we prepare better social responses to behavioral presets when we know why they exist and can therefore target interventions more effectively.

In other words, nothing in this theoretical framework suggests we throw up our hands about social reality and accept whatever is currently taking place as inevitable. We might however have to recognize some limits to the degree of change we can introduce without drastic unintended consequences. If the desired goal were to reduce rape to close to zero, regardless of other consequences, there are avenues of action that might get us there, or at least much closer than we are presently. You could, for example, simply chemically castrate every male who was ever even accused of a sexual assault and restrict the freedom of young men and young women in certain social settings—e.g., college partying and socializing environments—in draconian ways. The harms in such interventions would be substantial, of course, and we would need to carefully consider and weigh the costs and benefits. Few of us would likely be willing to take on the costs of the program for rape reduction just described. But having the knowledge of what in our evolutionary history is contributing to the phenomenon of murder or rape would certainly put us in a better situation to address such problems than the one we are in at present, where our knowledge about human nature and about why we engage in such behaviors as murder and rape is fundamentally skewed by unrealistic and exhaustively environmentalist theories of the causes of human behavior.

An evolutionary perspective focuses acutely on a central question: What makes us human? A reasonable social science must begin, at least axiomatically, with a stand on this question. Inevitably, this will get us into an effort to speak about human nature, even if the effort ends, as it commonly does for many mainstream sociologists today, in the denial of the very notion that there are any universal, pre-sociocultural characteristics of our species. But a properly scientific sociology must begin with the assumption that there is something we can reasonably call "human nature," though it be nothing more than the outline of characteristics that separate us in some way from other animals and legitimize the category "human" as a distinct category.

One of the central frames for talking about the nature of the human and what makes us what we are is the nature/nurture opposition. As a hermeneutic mechanism that makes outlining a course syllabus easier, the opposition has utility. Certainly, there are plenty of analytic ways to distinguish between elements of the innate makeup of the human being and the environmental forces of various types that act on him even before he is born. But the deep truth of the theory of natural selection is that we must assume that almost everything we might be interested in looking at in an organism is a product of the interaction between the innate and the environmental. Each of the two acts on the other in an endless complex dance of change and adaptation. In the words of one of the most important figures in modern evolutionary biology, Theodosius Dobzhansky, "[g]enetic traits are not preformed in the sex cells, but emerge in the course of development, where potentialities determined by the genes are realized in the process of development in certain environments" (Dobzhansky 1962:8). The challenge I sometimes hear from some of my students after they have been exposed in my class—likely for the first time—to sociobiological accounts of some human trait or aspect of social organization is 'But it can't all be genes and biology!?' No, it is not. But whoever said it was? The point is not to reject a one-sided environmentalism—the current mainstream face of sociology—and replace it with an equally implausible one-sided genetic determinism. It is to recognize that while the heuristic device is perhaps still unavoidable in some settings, the idea of a nature/nurture divide—as opposed to a nature/nurture symbiosis—has outlived its utility. We must move on to the delightful complexity of realizing that nearly every single aspect of human behavior and social organization that interests us is a result of the interplay between ecological or environmental factors that include both the natural and the social environments, on the one hand, and the basic biological blueprint of the organism we are, on the other.

As J.B.S. Haldane notes at the outset of his book *The Causes of Evolution*, evolution and natural selection are perfectly separable concepts. One can understand something as evolving, in some way, without accepting the actual theoretical engine for evolution that was suggested by Charles Darwin and then augmented subsequently after his death into what is now known as the neo-Darwinian synthesis. Theories of evolution that have nothing to do with the theory of natural selection are not hard to find in the history of sociology. Indeed, the basic progressive narrative that informs most of what makes up contemporary critical theoretical sociology is a sort of evolutionary perspective. According to this approach, in social and political terms, things get progressively better and better as activists concentrate their attention on first this form of oppression, now that one, until they are all addressed and solved, and the social utopia emerges.

A sociology based on the theory of natural selection as its fundamental theoretical framework is another thing altogether. Briefly, here are the basics. Since Darwin, the theory of natural selection has been augmented in many ways, and its increased complexity has only lent it more power to explain evolutionary process and result. But at its very most basic, the theory of natural selection can be summarized quite succinctly. It is an intellectual tool for explaining why organisms are constructed as they are in a given time and place that consists of three principles:

- phenotypic variation that is correlated to genotypic variation
- heritability of traits from parent to offspring
- fitness consequences of different phenotypes.

Phenotypic variation in the population means simply that the members of the population must not be identical copies of one another, and genes are an important causal force affecting phenotypes. Heritability occurs through sexual reproduction and recombination, which fundamentally shape the way in which traits are passed along to offspring as well as the tendencies of particular kinds of outcomes in mate pairing. Fitness consequences mean simply that the presence or absence of the traits must provide advantage or disadvantage to individuals in reproductive competition with others. It is ecologies themselves that "select," if you will, among different organisms insofar as they provide different combinations of advantages and disadvantages to those organisms in the business of survival and reproduction, and some organisms have a greater balance of advantages than others. There is a basic, brute process underway in any ecology in which the conditions reward those organisms with phenotypes that most advantage them in that ecology by making it easier for them to achieve the difficult tasks of survival and reproduction. The genes that have produced the successful phenotypes are passed along to offspring, who will then tend to have the same favorable phenotypes and genotypes, and over time these phenotypes will outcompete others and come to be more numerous in the population. Mutation is a constant source of innovation in the genomic material, but most of it is either of neutral impact on fitness or it is deleterious.

The theory of natural selection operates fundamentally at the level of the individual organism, or indeed even further down in the hierarchy of the organization of life, that is, at the level of genes, chromosomes, and genomes (Alexander 1987:37). The discussion of possible group selection continues in evolutionary biology circles, and the debate between advocates of individual-level selection and group-level selection is often intellectually rich. Some biologists such as Stephen Stearns have suggested that our species might currently be in the process of a transition from individual selection to group selection that has stalled because of the counteracting forces of individual human biology and culture (Stearns 2007). The individual organism is ultimately, in the language of the modern neo-Darwinian synthesis, a collection of genes under the pressure of an ecology, and since no two individuals have precisely the same genes, conflicts of interest between individuals are expected to be ubiquitous. Selection is most viable in humans at the individual level, and even when it might reasonably be said to operate at the group level, this is because individuals inside groups are advantaged by their group doing well in competition with other groups. Group on group antagonism is one of the chief threats and drivers of evolution in our species, given the fact that tens of thousands of years ago we achieved the remarkable success, thanks to our group life and technological advance, of basically eliminating the previously endemic threats from predation.

Individual organismal interest has to be understood as reproductive and therefore—given sexual reproduction—inclusive of the interests of closely related others, since any individual shares at least some of his genome with others and can therefore be expected in at least some scenarios to benefit personally from the reproductive success of those genetically related others. In this framework, interest is always selfish, in the inclusive sense, and all behavior that appears at any level to be—or that is understood by the actor or others as—selfless, sacrificial, or altruistic should be closely theoretically interrogated. Nonetheless, talking of interests as selfish does not mean that evolution is in some simple and straightforward way about survival, either of human groups or individuals. Only genes are evolved to survive, and they can do so abundantly even if given individuals perish. Any given individual may, depending on the situation, have a better chance of reproductive success through an act that brings about his own demise if that act also ensures the reproductive fitness of closely genetically related others. Evolution has produced all life, including us, to serve the interests of the genes that inhabit us. Conscious, reflective life such as *Homo sapiens*, however, which is now even capable of accurately understanding the forces that drive us for the first time, is not obliged to serve those interests, and so things have become quite complicated indeed at the human level (Alexander 1987:40).

It is nonetheless an empirical question, not one to be settled *a priori* by our desires about what kind of creatures we would ideally like to be, as to whether any given behavior or social practice can most effectively be made sense of within the framework of inclusive fitness. Relatively recent developments in evolutionary theory make it possible to understand in this manner a great deal of behavior that previously was considered unamenable to explanation by a Darwinian model. Kin selection is the process by which individuals might be expected to pursue inclusive reproductive interests in ways that appear to be self-sacrificial when the beneficiaries of the action are close genetic relatives. The parent who willingly suffers or deprives himself for a child is the typical example, but the logic extends to other degrees of relation as well, e.g., siblings, nieces, and nephews. Reciprocal altruism is the theoretical innovation developed to understand how behavior that seems self-sacrificial and benefits someone other than a close relative might still make most sense in an inclusive fitness model because of social rules of reciprocity and status systems that grow up alongside them. I will have more to say about this in a discussion of reciprocity and morality in chapter four.

Studying human interests as fundamentally reproductive means uncovering all the ways in which human life is driven ultimately by the production and/or promotion of offspring and proximately by the various behaviors and beliefs that could be expected to further that in the human natural environment. Not all this behavior, but some significant portion of it, will of course involve ideas and practices related to the basic act of sexual reproduction itself. This is a topic that is exceedingly difficult to study accurately, not least because people are often highly interested in deceiving others about this sphere of their own behavior and thinking. Data sources are therefore often suspect and weak. Donald Symons suggests that autobiographical documents like the book *My Secret Life*, the erotic diary of a British man who had hundreds, if not thousands of sexual interactions and documented them fairly extensively, are perhaps the best we can do in the way of objective accounting here. Survey data cannot hope to fully avoid problems of validity, but we cannot avoid using them (Symons 1979:66–74).

What is the evidence that human behavior is only fundamentally understandable by reference to reproductive interests? Answering this question requires beginning with an inquiry into what we mean when we talk about human interests. At an only slightly oversimplified level, we might talk about the modern social sciences as a systematic effort to confuse the question of human interests by ignoring evolutionary biology and proliferating a hyper-simplistic plurality of theories of interest, each incompatible with the other and all seeking intellectual dominance over competitors. Political science—the oldest of the social sciences—sees power as the fundamental interest pursued by humans. For economics, it is material wealth. Psychologists argue for personal contentedness. Sociologists believe that the fundamental interest driving human beings is the desire for social harmony and community. None of these theories of interest withstand serious critical interrogation. Unquestionably, many people much of the time do indeed seek power, financial gain, happiness, and fellowship, but history is also full of examples of human action that cannot be fitted to any of these theories. Political leaders walk away from open avenues to empire, the wealthy give away their riches, happiness is indefinitely deferred and in some cases even rejected outright by individuals who spend their lives in stressful pursuit of something they might believe will eventually make them happy but that never does, and people flee from their families, friends, and countrymen to live in the woods and avoid contact with other humans, or walk around on busy city streets totally isolated in their smart phone-mediated personal worlds. The test is ultimately empirical. What view of interest is most likely to have the most explanatory purview in the rich array of human behaviors and ways of organizing themselves socially?

A problem that has to be admitted at the outset for the theory of reproductive interests stemming from the Darwinian evolutionary framework is that behavior can only be determined to be adaptive when we consider it with respect to the environment that is its natural environment, that is, the one in which it evolved. Since humans no longer live in their natural environment—that of hunter–gatherer social orders on the sub-Saharan African savanna, during the latter part of the Pleistocene era that stretched from around two and a half million to 12,000 years ago—determining the adaptiveness and the ultimate causes of behaviors will be complex, difficult tasks. A natural selection view of human interests may also require us to augment, or even fundamentally alter, presuppositions of a philosophical nature about the context and cause of human action. If behaviors are a consequence of selfish genes being selected by natural selection, where does our autonomous choice enter in? Can we be said to act according to our free wills, in any traditional philosophical sense?

The biologist Richard Alexander developed a compelling framework for understanding human interests over the course of the lifespan. A human lifetime is, in this view, understood as a changing calculus of the balance of somatic, parental, mating, and extra-parental nepotistic interests. At the beginning of life, all interest is somatic. We try to increase ourselves as organisms, to grow and survive long enough to reach the reproductive years and be presented with opportunities to reproduce. As we enter the reproductive years and the middle of life, somatic interest starts to steadily decrease, and mating and parental interest increase. Then toward the end of life, all varieties of interest except extra-parental nepotistic interests decrease (Alexander 1987:41). Lifetimes, Alexander argues, are finite because of pleiotropy, that is, because many genes do more than one thing at different stages of life. This means that any gene with early beneficial effects and late deleterious effects, or early beneficial effects and no effects later, would tend to push toward aging and senescence unless it could be replaced by an allele with beneficial effects throughout the life cycle, which is unlikely. At the time of the writing of Alexander's book in the 1980s, not much was known about specific genes and pleiotropy, but we now have many studies that back this up (Leroia, Bartkeb et al. 2005; Carter and Nguyen 2011; Byars, Huang, Gray et al. 2017). The very existence of senescence is evidence of the accuracy of the Darwinian argument that human interests are reproductive at their core. Adaptive success is not calculated merely by the longevity of an organism, but rather by its success at successfully reproducing copies of its genes.

Interests and the conflicts among them then produce four basic types of human relationship, defined by the degree of common interest and the concomitant balance of honesty and deception that characterize the relationship and its constitutive interaction. In the situation of a total overlap of interests between A and B, their relationship would be characterized by no deception and total transparency. By the logic of how interests work in our species, no actual human relationships ever attain this state. The second type is restrained embellishment, where there is a substantial but imperfect overlap of interests, as in the case of a parent–offspring, sibling–sibling, or bonded mate relationship. Here, the degree of truth in the relationship is considerable, but there is nonetheless room also for deception, e.g., extra-marital cheating or betrayal of a sibling to achieve a better take of a parental will. Unrestrained embellishment is characteristic of relationships between those who are not closely related and are at least with some frequency directly in competition of some intensity, e.g., competitors of the same sex in a dating and mating market. There is here a small core of truth, but deception is significant, limited fundamentally only by the ability of competitors to detect deception and thereby effectively counteract it. The fourth type is characterized by interacting individuals with no overlap of interests at all, and here deception characterizes all behavior. Alexander gives predator–prey relations as exemplary here, and so as in the first type we can expect few or no real human interactions to exist here. It is in the middle two types, then, that we see the vast bulk of human social life taking place (Alexander 1987:74).

On the Non-Darwinian Mode of Thinking Known as Social Darwinism

Those averse to thinking in evolutionary terms about humankind in academic settings often know little of the history of evolutionary thinking, but they have almost always heard something about Social Darwinism, and they are not infrequently quick to invoke it as justification for dismissing evolutionary thinking in the social sciences altogether. The history and legacy of Social Darwinism is a complicated topic I do not have the space to consider in detail. The short version of a much longer story is that it warped Darwinian selection in ways that are not at all consistent with the most basic elements of the theory. Spencer's formulation of "survival of the fittest" easily lends itself to the narrative of historical progress that Spencer accepted and that Darwin did not. "Fittest" should be understood in the accurate Darwinian view as "fittest among the competitors in a given environment at a given time." Natural selection is not about an inevitable process of churning out of the ideally fit. An organism need only be, as Dobzhansky et al. (1977) write, "tolerably fit" to outcompete others less well fit. If the ecology changes, however, what natural selection will determine to be tolerably fit may well shift as well. There is no moralizing, progressivist narrative to be found in the theory of natural selection. Its logic is not perfectionist; it produces what works well enough, from the material available. This almost always means significant tradeoffs that a perfectionist adaptive mechanism would never produce. A single gene, if passed along in the appropriate allele by both parents, produces cystic fibrosis, a lethal disease. But the same allele that produces the disease also, in single form inherited only from one parent, provides protection against typhoid fever. When the risk of typhoid is significant enough in a given environment, natural selection can be expected not to remove the allele that confers protection against it from the population, even if the consequences are less than perfect (Hrdy 1999:107). The same logic is at play in the existence of the sickle-cell anemia gene, which in double form can be lethal but in single form confers protection against malaria.

Social Darwinism sometimes oversimplified the logic of fitness and the reality of adaptation in ways consonant with a certain political ideology. There is no one stable, unchanging set of behavioral predilections that will certainly produce fitness in all or even in most environments. Size and strength are typically advantageous in some ways—increased ability to intimidate or physically overpower rivals and attract mates—but disadvantageous in others—less speed and agility in escaping from predators, need for more food to maintain bulk and muscle. Stealth and deception are underappreciated adaptive traits. Smaller male cuttlefish sometimes achieve mating opportunities by deceiving larger males who might otherwise chase them off—they do this by camouflaging themselves as females. Add human culture to this game and things get quite a lot more complicated. Social Darwinism too frequently asserted a "one size fits all" model of fitness. Some of its proponents also insinuated that given phenotypes in human populations were essentially stable and innate to certain populations. Racist philosophies sometimes undergirded such insinuations, whether implicitly or explicitly. But the theory of natural selection takes any given phenotype not as an unchanging category, but as a product of evolutionary history that is always in flux informed by the stimulus of the environment and subject also to chance in the form of mutation. The explanatory framework works in such a way that given sufficient time and the right ecological context any given phenotype can be altered so radically as to make it unrecognizable. Given the right ecology and enough time—many millions of years—you could see a microbe transformed into a Sequoia tree. Indeed, this is exactly what we believe happened in evolutionary history. In our species, there can be no metaphysical arguments about the superiority of this or that sex or this or that race in natural selection, and an attempt to use it to make such cases shows a basic misunderstanding of the theory's premises.

Before closing this brief parenthesis, let us note that it was not only the Social Darwinists a century ago who brought politics illicitly into the discussion of whether evolutionary theory could be made useful to sociology. Some of the most ardent critics of this intellectual move within the community of scientists have been at least as egregiously polemical and partisan in crude political terms. In a marvelously insightful work of intellectual history, Ullica Segerstråle sketches out the history of early debates over sociobiology—initiated by the publication of E.O. Wilson's massive text on the topic in 1975–at some length. Richard Lewontin and Stephen Jay Gould, the two figures in biology who are consistently trotted out in mainstream sociological dismissals of sociobiology and evolutionary theoretical applications to the social sciences, were quite evidently driven to some significant degree by their radical left-wing politics in their interventions on this topic, and they behaved themselves quite frequently in an intellectually questionable manner. The Harvard biologist Bernard Davis criticized the attack on Wilson by Gould and Lewontin as "Neo-Lysenkoism." He wrote a crushing review of Gould's *The Mismeasure of Man* in which he noted how negatively the book was reviewed in the scientific community, while it was being fêted by activists and non-scientists, including many sociologists. Lewontin wrote an entire book, *The Dialectical Biologist*, attempting to square his professional discipline with a revolutionary Marxist politics, and he even suggested—in a manner that is perfectly consistent with the core of Marxism—that he considered scientific truth at the end of the day to be subservient to political ends:

> Of course the speed of light is the same under socialism and capitalism, and the apple that was said to have fallen on the Master of the Mint in 1664 would have struck his Labor Party successor three-hundred years later with equal force. But whether the cause of tuberculosis is said to be a bacillus or the capitalist exploitation of workers, whether the death rate from cancer is best reduced by studying oncogenes or by seizing control of factories—these questions can be decided objectively only within the framework of certain sociopolitical assumptions...science and other aspects of social life interpenetrate and...scientists, whether they realize it or not, always choose sides.
>
> (Levins and Lewontin 1985:5)

A central argument in the book is that the elimination of capitalism is a requirement for achievement of the true dialectical science Lewontin champions (ibid.:287). He and Gould were central members of the Sociobiology Study Group, a group of radical academics created to combat Wilson and affiliated with an explicitly Marxist group that called itself "Science for the People" (Segerstråle 2001:18–24). Lewontin was also apparently the individual behind the pseudonymous Isadore Nabi, a fictional scientist who was allowed in a number of scientific venues to respond to Wilson in a way that can only be described as trolling before the fact, with the complicity of other intellectuals hostile to Wilson (ibid.:184–188).

The Yale biologist Stephen Stearns, writing at the death of Gould, summarized his contribution to thinking on evolutionary theory in the following terms:

> I now frequently encounter biology undergraduates, nonacademics, and even some nonevolutionary biologists who have read Gould and carry quite alarming misapprehensions about natural selection, adaptation, and microevolutionary mechanisms, but insist that they have understood evolution at the hand of a master and need go no further.
>
> (Stearns 2002:2345)

Stearns described Gould as a popularizing "Man of the Left," comparing the response to his death in the major media to the reaction the passing of the French activist and writer Victor Hugo provoked. If we are going to point out the egregious way in which politics intrude into science with Social Darwinism, as we should, we are required to do the same with respect to the radical left intellectual enemies of Social Darwinism who were equally incapable of keeping their politics from warping their view of the science.

Genomics and Social Theory

I am unaware of any reliable survey data on this, but my anecdotal sense from two decades living professionally inside one social science discipline (sociology) and interacting with some frequency with many colleagues in other social science disciplines (economics, political science, anthropology) is that the dominant position in the social sciences on the contribution of genes to human behavior is that it is minimal, and perhaps nothing at all. There are some— I have met them in the course of a quarter-century working in universities— who take this position to its logical extremes given the doctrine of social construction: That is, they literally do not think that knowledge in behavioral genetics and related fields constitutes anything more than the political druthers of the individuals who make up those fields. Some of these figures have astonishingly elevated platforms and large audiences in the academic community. More common, though, is the perspective that was communicated to me once by a colleague when we were discussing whether or not it was likely we would ever be able to link specific genes to given behavioral predilections in any kind of systematic way. "No," my colleague said, "that won't happen because behaviors are too complicated. Genes can cause certain diseases and simple physical features, but behaviors are just too complex."

But my colleague's dichotomy will simply not hold. First, in its simpler forms, it relies on the notion that disease is some relatively uncomplicated state of the organism, while behavioral predilections are greatly more complex matters that cannot be mapped in this binary way. The truth, as Robert Plomin (2018) has argued, is that much genetic disease is better thought of not as "on" or "off" conditions, but as gradations between two poles, a spectrum, if you will. We know, for example, not only that, contrary to the psychological orthodoxy of a half-century ago, schizophrenia is not caused by the childhood environment, but that it is best characterized in more or less the same way that we characterize any other complex personality trait such as, for example, openness to experience. There is no schizophrenia gene. Those who have the condition differ genetically in quantitative terms from those who do not. They simply have a larger number of the genetic variants that nearly all of us have, a sufficient number of which begin gradually to produce the psychologically pathological condition we name schizophrenia. The same is true for many other conditions we categorize as "you've got it, or you don't" pathologies, e.g., depression. If this is true of much mental disease, why would we imagine without evidence that the same thing is not true of other mental attributes?

It is of course true that some disease states are linked to single or small numbers of genes. The same is true for some behavioral predispositions that we would not typically classify as diseases. There is for example an array of genes under research as linked to exaggerated predilection to violence and anti-social behavior. The most well-known is the example I briefly mentioned in the preface, the gene MAOA, which normally makes an enzyme, monoamine oxidase A, that metabolizes neurotransmitters important for impulse control. In forms of the gene that produce low levels of the enzyme, higher levels of violent behavior are typically found, especially when combined with the appropriate environmental triggers. Beyond this gene, a number of others that are implicated in the production of serotonin and dopamine—neurotransmitters associated with impulse control and helping us learn through experiences of pleasure and pain which actions to repeat and avoid, respectively—and that thereby affect propensity to violence are being studied, including 5HTT, DRD2, DAT1, and DRD4 (Raine 2013:56). The mechanisms and relationships involved are highly complex, and careful critics rightly point out that we must be cautious in too easily positing genetic causation in situations in which it is clear that nature and nurture are—as they essentially always do—working together to produce specific behavioral predilections. Biological risk factors for sociopathy, for example, include certain genomic states, but they also include neurological deficits that are connected to environment, e.g., exposure to lead as a child. And frequently it is the case that allelic states of particular genes alone are insufficient to produce substantial rises in sociopathic behavior. For sociopathy to occur, the individuals affected must be subjected to negative familial situations of neglect or abuse as well. Facile claims that e.g., testosterone levels alone can predict levels of aggressivity have been shown to be too simple—it is the balance of several neurochemical factors tied to genes that are at work here. Testosterone alone might help us understand the gap on aggressivity between males and females, but when we want to explore different levels of such activity among males, we need to add in the levels of stress hormones such as cortisol and neuromodulators such as serotonin. High testosterone, low cortisol, low serotonin individuals are more likely than others to be physically prone to aggression with poor impulse control, while men with high testosterone, low cortisol, and high serotonin might show similar physiological aggressivity indicators but be better able to control impulses (Churchland 2013:146).

And what about the way in which phenotypical characteristics that are presented by simple mutations at a single gene interact with culture and therefore with behavior in intricate and fascinating ways? The effect here is not causal in any simple way, but the connection between the genetic level and the cultural level certainly ought to be a normal element of sociological inquiry. The gene that codes for lactose tolerance, a mutation of the original genetic programming in humans that is still present in most of the world's population in which the protein that allows the digestion of lactose stops being made in childhood, does not drive the specific cultural place occupied by dairy-based foods in the US or France—I often talk about the 'La Vache qui rit/The Laughing Cow' cheese trademark in discussing this with students–in any specific direction. But it is no misstatement to say that it is literally the cause of those cultures, since without lactose tolerance in a significant number of adults in any society, there would be no widespread use of milk-based foods in given populations and hence no cultural frameworks for making meaning out of that use.

Sociologists and others ignorant of the state of the research in behavioral genetics greatly underestimate the effect of heritability on the full panoply of behavioral attributes in humans. The psychologist Eric Turkheimer, in a now classic 2000 paper, provided a list of three basic laws of human genetics regarding behavior: 1) all behavioral traits are heritable to some degree, 2) the effect of genes on behavioral traits is greater than the effect of growing up in the same family, and 3) there is a substantial portion of the variation in behavioral traits in humans that is not predicted by either genes or familial environment (Turkheimer 2000). The most basic psychological attributes of an individual human are heritable, many to quite a significant degree. For example, all the Big Five personality traits—extraversion, agreeableness, conscientiousness, neuroticism and openness for experience—are highly heritable, genetics explaining between 40% and 60% of the population variation in these traits (Power and Pluess 2015). Verbal ability, spatial ability, and—as I will explore more below—general intelligence are all heritable and the variation in the population on these traits caused by genes ranges from 50% to 70% (Plomin 2018:6).

The basic fact of the way in which genes affect behavioral predilections complicates the way in which sociology must build behavioral genetics into its theoretical framework. Unlike in some diseases, where a single gene is responsible for the difference between disease and non-disease state, genes that affect behavior nearly always do so in a polygenetic fashion. That is, the presence of one allele, or even a few, is not going to have a significant effect on complex behavioral or personality traits. It is combinations of a larger number of genes that do this work. For this reason, methodological problems plagued early attempts to study the effects of genes on behavior and personality. Sample sizes for studies were typically small, and these studies were of small numbers of preselected or candidate genes, not on the entire genome or even significant large sections of it. False associations were a result of much of this early research. Behavioral genetics today is based on GWAS, or genome wide association studies, which are methodologically much richer. Instead of looking only at a small number of candidate genes, GWAS concentrate on large numbers of genes and compare how SNPs, or single nucleotide polymorphisms, affect different organismal states. SNPs are the most common way in which individuals differ genetically, *via* variations in single nucleotides, which are the fundamental building blocks of DNA. Individual humans have about 4 million SNPs in their genome, and there may be a total of around 80 million SNPs in the world (Plomin 2018:113). GWAS allow you to search the entire genome for as many as a few hundred thousand SNPs, which gives you a much more quantitatively deep dive into the possible correlations of genomic structure and given behavioral variables (ibid.:119).

There is some debate about how much even GWAS will allow us to see concerning behavioral predilections. The neuroendocrinologist Robert Sapolsky, for example, is at the pessimistic end on the question of whether such genomic research will ever allow us to understand with any clarity the causal effect of genes on behavior. A common complaint is that the quantitative aspect of the work is just too unwieldy. Others, however, point to the fact that we are adding magnitudes of power to our methodological toolkit every year, and when we take an historical approach and look at where we were, say, just a decade ago, there is no reason to believe that any of the methodological difficulties are insuperable. This writer finds it helpful to consult the history of science literature to help pick a side on this debate. The list of scientific disputes in the past on which at a given time A at least some high-profile figures in science were predicting an insuperable difficulty had been reached and then, at a later time B, the insuperable was achieved is quite a long and impressive list. I would not bet against significant scientific advance on this front.

Genomic information, specifically the relationship of polygenic scores to behavioral predilections and personality traits, will be relevant to a great deal of the research topics in which sociologists are interested. Charles Murray offers a useful insight into just how much sociological research will be affected by noting how many *American Sociological Review* articles in 2018 and the first issue of 2019 dealt with just such topics: fully two-thirds of them, or 33 of 49 (Murray 2020:288). As he puts it, "[t]here's no longer any question whether the use of polygenic scores will be widespread" (ibid.:293). Genomic data alone is now proving capable even of showing us insights into social structures in societies that have long since ceased to exist. Recently, DNA remnants were analyzed from a small Neolithic group of individuals that had been collectively massacred and buried together. Researchers were able to compare the genomes of these individuals to those of currently existing humans and determine that modern Sardinians are the existing human group genetically closest to the last Neolithic populations. They were also able to demonstrate with the DNA data that the group in question must have practiced patrilocalism, which is the form of marital system in which males stay in natal groups and females migrate to the family group of their husbands, because the males in this burial site were all of the same Y chromosome group, i.e., closely related, while females were much less closely genetically related (Schroedera, Margaryan, and Szmyt 2019).

Sociologists are still far behind the curve in accounting for this intellectual revolution, but there are here and there exciting exceptions. The University of North Carolina at Chapel Hill, for example, has a small group of scholars in a concentration on Genetics and Society, led by Guang Guo. This is the same department that once housed Gerhard Lenski and François Nielsen, both of whom studied inequality from an evolutionary-ecological perspective. Scholars such as Guo are at the forefront of the substantive effort to inject genomic data into the study of traditional sociological topics such as deviance and antisocial behavior. He and his collaborators have contributed a stream of articles on, among other topics, the effects of marriage on antisocial behavior in those with significant genetic risk to such behavior (Li, Liu, and Guo 2015) and the interplay of peer influence and genetic predisposition to alcohol abuse on alcohol use (Guo, Li et al. 2015).

Guo was the lead author of an article published in a landmark event for genomic sociology, the *American Journal of Sociology* special issue in 2008 on genetics in sociology. It was, as it turned out, the only piece in the volume that found that genes seemed to have a substantial effect on the variable studied, a telling statement about sociology's preparedness to fully acknowledge the impact of behavioral genetics. Indeed, the guest editor of the special issue, Peter Bearman, made clear the general drift of the discipline in introducing it by opposing two reactions to research on genetics and human behavior—"enthusiastic embrace and uncritical adoption (as represented in the *Science Times*) and fear and loathing (from sociologists)"—and then editorializing that "one realizes that they arise from the same source: a naïve overvaluation of 'genetics'" (Bearman 2008:v).

In this *AJS* article (Guo, Tong, and Cai 2008), we encounter an exemplary vision of how genomic data can be utilized in sociology. The bulk of this work, Guo and his co-authors argue, will require GWAS investigation of large numbers of genes that individually have only small effects on the behavior in question but that collectively exercise considerably more influence. But this does not mean that smaller genetic units cannot be fruitfully included in sociological studies of behavioral predilection. Guo and his co-authors looked at a single gene, DAT1, that codes for a protein involved in regulating the production of dopamine, a neurotransmitter known to play a significant role in male but not female sexual arousal and behavior. They found that individuals with one, relatively rare allele of the gene show significantly fewer sexual partners than those of the most commonly found other alleles. More, individuals with the rare allele were generally more risk-averse and typically scored higher in intelligence. They examined the effects contributed by several important social contextual factors—socioeconomic status, religious activity, school environment—and found that only the last of these significantly modified the effect of the DAT1 gene. Individuals with the rare allele were less affected by its influence when they attended schools in which levels of sexual activity were high, that is, peer pressure was apparently sufficiently strong to counteract the genetic factor. Their conclusion is measured, indicating the intention to attempt to replicate the study with a larger data set when such is available. The crux of their finding, they reiterate, is "the complex interplay between socioeconomic–cultural influences and individual genetic predispositions" (ibid.:S61).

Twelve years on from the publication of this relatively early sociogenomic research, the data on the role of genes in myriad behavioral variables is overwhelming and undeniable. Yet as of 2020, there is only one ASA section that has anything to do with evolutionary theory, genomics, or neuroscience. The section on Evolution, Biology, and Society is the single smallest section in the ASA body of about 50 total sections, with a membership of just over 100. The student membership is only 20, indicating a poor future for the section. By contrast, the highly politicized and scientifically anemic sections on Sex and Gender and Race, Class, and Gender have nearly 400 and 350 student members, respectively.

In 2018, *The Genome Factor*, a book on social genomics by two sociologists at prestigious universities, was the focus of some attention in the discipline. Careful readers of the book, though, will be unable to avoid the impression that it is as much an effort to limit the role of genetics in sociology as it is to investigate its promise for sociological research. The book's authors, Dalton Conley and Jason Fletcher, spend a good deal of time insinuating that the methodological complexities of untangling genetic causation in behaviors—the need to analyze large numbers of genes in sizable samples, for example—may well prove insurmountable. They show a particular allergy for any insinuation that there might be genetic components in the working out of social hierarchies and inequality that may make it relatively difficult to alter those features of social order through traditionally-oriented social engineering efforts. The book also concerns itself at some length with debunking a simplified caricature of the claims made in *The Bell Curve* about the genetic role in class inequality. At a 2018 ESS session on the book, Conley described his reasoning for pursuing a degree in genetics after working for some time as a professional sociologist as rooted in his desire to challenge claims that were being made about the significant genetic contribution to inequality, on which he had already written a good deal from a more or less conventional, environmentalist perspective. One reads *The Genome Facto*r as at best a grudging admission that genomic data are here to stay, and good chunks of it show no real connection to a social science based in the neo-Darwinian synthesis.

The book's lukewarm embrace of genomics and steadfast defense of much of the mainstream discipline's antagonism to deep evolutionary thinking has not apparently been enough to persuade sociologists to pay more attention to the topic. At the 2018 meetings of the ESS, I attended both an author meets critics session for *The Genome Factor* and a panel on genomics in sociology. Neither attracted many conference attendees.

The Brain as Evolved Object, Heritability of Cognitive Ability, and Sociology

In her introduction to a neuroscientific way of reconceiving human identity, Patricia Churchland points in the second sentence of the book to the core truth that must be recognized by sociology: "I am who I am because my brain is what it is" (2013:11). If that is true, then everything that has to do with the relations that any individual human individual involves himself in are also what they are because of what their brains are. What would a sociology that took the brain seriously look like?

A human being is of course not literally a brain. A brain requires a body to act. All that we are learning about how brains operate show us that in fact they require yet another thing to act, and even to developmentally become the kind of things capable of acting: other brains. The brain is a fully social entity, in this sense. And though a person is not literally his brain, it is nonetheless true that everything he can do as a human being—have thoughts and beliefs, strategize and reason, feel emotion, act in the world in concert or conflict with others like him—requires a functioning brain and is impossible without one.

As David Franks and Jonathan Turner argue in their introduction to their edited handbook on neurosociology, the discipline of sociology will either mine the rapidly growing body of data on the ways in which the neurochemical state of the brain affects behavior and interaction or it will be increasingly irrelevant in the intellectual task of fully mapping the causes of human social behaviors. The rich subfield of the sociology of emotions has, in its most scientifically rigorous incarnations, helped renovate the theoretical model of human consciousness and decision-making to accommodate what we know now about the primacy of emotional experience over purely rational cognition in driving human behavior. Neurosociology promises to tie these insights into experimental data on precisely what processes in the brain drive emotional affect in reaction to stimuli in the world, including stimuli that come from human interlocutors in social settings. I will have more to say on the importance of the brain's role in producing emotional experience for the understanding of human evolutionary history and human nature in the next chapter.

Though the priority of place of cognitive function in brain evolution is being reevaluated in light of recent research, the brain's role in shaping an individual's life chances and likely trajectory in society as a direct result of its capacity for rational problem-solving remains an important topic for social scientists to study. Yet Stuart Ritchie begins his short summary of the research on intelligence, or IQ with the profoundly accurate sentence "Smart people don't like the idea of intelligence" (Ritchie 2016). Sociologists have been especially allergic to the topic, sometimes denying it exists altogether, then claiming that it is merely the artifact of particular social systems based in hierarchy and inequality, then trying to turn it from one unified concept into a democratic plurality of different personality traits and abilities in which virtually everyone can rank relatively high on some kind of "intelligence" even if she fares less well on others.

In scientific communities that study intelligence, though, there is no significant debate about the existence of the concept. It is an empirically verified category by which individuals can be objectively scored. Those who reject the category "intelligence" as something that does not exist, or is not accurately verified by IQ tests, or who perhaps admit that intelligence exists but believe that current IQ testing is too hopelessly flawed by cultural bias to accurately detect it have an uphill battle indeed to convince those concerned with data, given the state of the research. In 1994, the psychologist Linda Gottfredson and 51 co-signatories published in the *Wall Street Journal* a statement on "Mainstream Science on Intelligence" intended to set the record straight on what most experts on intelligence thought about the state of the scientific research on the topic in the wake of the public turmoil that followed the publication of *The Bell Curve*. Already a quarter-century ago, the clear consensus among such experts was that IQ is a fairly coherent, well-defined, and unitary category of psychological science, with a deep body of research supporting its validity as a concept, and that it can be accurately measured in all human groups with current tests. In the intervening twenty-five years, that consensus has only gotten more solid. Contrary to a claim that is frequently made, corrections have long been made for cultural or experiential biases in previous forms of intelligence tests. Individual levels of intelligence tend to distribute along a normal curve, and the same is true for groups as defined by conventional racial identity. Gottfredson even demonstrated in her later discussion of the background of the statement that only a very small minority—a mere 7%—of the more than one hundred scholars she asked to consider joining the statement responded by stating they would not do so because in their view it did not represent the mainstream science (Gottfredson 1997).

Intelligence is a broad mental ability that permits individuals to effectively answer questions and solve problems. It is not uniformly distributed through the human population, that is, some people have more of it than others. Learning is certainly an important element that contributes to an individual's intelligence, but there is also a significant genetic component. Contemporary research indicates that about half of the variation in IQ is heritable, i.e., genetic, and this fluctuates over the life course. In children, about 20% of the variation is genetic, while in adults it can be as much as 80%.

Much early criticism of the genetic contribution to IQ argued that one environmental influence—parents—must have an outsized role to play. But we know now that much of the parental influence that had previously been assumed to be about the parents' economic, social, and cultural contribution in fact seems to have to do with genetics instead. What parents do with children has much less effect on children's IQ than mainstream sociology typically assumes (Harris 1998). Outside of extreme cases of neglect and abuse, parents as an environmental effect have significantly more limited ability to shape children's personalities, including their IQ levels, than is widely believed. It turns out that peers are a much greater general environmental variable in the life of children.

Research on the specific genes affecting intelligence is still in the relatively early stages, but already hundreds of genes related to intelligence have been found and are being further studied (Gouriounova and Mansvelder 2019). Two new loci in genes on chromosome 1 and chromosome 2, for example, were seen in a GWAS study of more than 35,000 individuals to be highly correlated with cognitive ability (Trampush et al. 2017). Typically, as in other complex traits, the individual effects of single genes on IQ are small, and the interaction between genes and environment is clearly of central import here. Nonetheless, we are in the process of empirically demonstrating that there is a range of likely profiles in cognitive aptitude in humans, and genes are the material basis of the range limits for given individuals.

Differences between individuals in IQ are indisputable as an aspect of human diversity. The question of possible heritable group differences in IQ is more controversial. Empirically speaking, we find evidence of group IQ differences along conventional racial category lines, but we do not know with precision yet what the causes of such differences are (Hunt and Carlson 2007). There are a range of positions taken in response to these data, some more tenable than others. One can accept that the group differences are real and then take a stance on where the cause is located, i.e., in environment or in genetic heritability or some complex combination of the two. Alternatively, one can reject the very category of "race" as a social construction, not useful for discerning and categorizing human differences. I will return at length to this topic in chapter seven, but suffice it to say here that there is considerable evidence that group-level genetic differences in different populations with different regional evolutionary histories exist. Given that fact, it is at least scientifically defensible to inquire into what if any differences in the general range of psychological and behavioral profiles might be associated with these populational genetic differences.

The scientific study of race group differences in intelligence requires attention to both possible environmental and hereditary contributions to those differences. Scientific research can be used in many different policy directions, that is, no determinate end is inevitable, whatever activists on this question might suggest. The quality of research, always important, takes on still more weight on this topic, but we cannot allow the daunting task of the difficulty of the work to prevent us from undertaking it. We made tremendous strides in the attempt to reduce race-based gaps in cognitive performance in the last half of the twentieth century, but now the gains seem to have stopped, or even to have reversed to some degree. Proponents of early intervention educational programs such as the Perry Preschool Project, the Carolina Abecedarian Project, and Head Start still make ambitious claims for their ability to produce robust gains in intelligence in children from disadvantaged environments, but the best read of the complicated available data on this indicates that any gains are typically quite modest. Long-term follow-up studies of individuals who experienced such early intervention programs show that apparently early gains in IQ fade over time, suggesting the superior causal influence of a hereditary baseline. While some positive advantages do seem to accrue to beneficiaries of such programs, they have been teased out as gains in "soft skills" such as motivation, conscientiousness, and self-control rather than as upward shifts in IQ. The claim that the remaining racial gaps in IQ are likely to be significantly modifiable with more interventionist programs along the lines of previous poverty programs is contested by much data. At least some portion of the remaining element of the gap due to environmental factors may well have to do with cultural differences between groups that will present formidable new challenges at the level of public policy.

Suppose, though, that the research of the past half century on this question is indicative of the nature of the group differences, and it turns out that roughly half the variation on IQ among racial groups too is due to genetically heritable factors. If we are interested in ameliorating this, is it not imperative to know as much as possible about precisely how the biology on this works, i.e., which genes are involved and how best to knock them out and replace with different alleles? Would this possibility alone not justify continued research into this question? And if the difference turns out to be unamenable to genetic engineering, or if as a society we decide against that for moral reasons, are we not better prepared to make our way forward as a society, to plan for alternative interventions or to recognize the degree of intractability of the problem and to discuss it openly, honestly, and publicly with that in mind, armed with the truth rather than willfully ignorant of the facts? Arguably, a multiethnic society like that of the United States needs to know such things in order to best devise ways to ensure national solidarity across ethnic and racial difference into the future.

The social policy implications for scientific knowledge of IQ are profound. There are almost always built-in assumptions of human nature in various policy programs. If the view of human nature assumed by policy is significantly out of touch with what science shows us to be so, we are at best using limited resources sub-optimally. We might be doing damage to the social fabric we do not even clearly see. As Edward Banfield (1974) once noted with respect to urban poverty programs, an endless repetition of "more spending on poverty will inevitably eliminate it" can have dangerous feedback effects if the claim turns out to be demonstrably false. Can citizens who are told such things by authorities be expected to remain enthusiastically supportive of mainstream institutions if it becomes clear that those authorities have misled them? Here's a real example: The No Child Left Behind Act—tellingly, subsequently renamed the Every Student Succeeds Act and renovated to remove national standards in light of the failure to achieve them—demanded that group-level test score gaps be reduced, but it prohibited giving different populations different instruction because that was determined as prejudicial given the Act's assumption that there can be no group-level differences in innate cognitive abilities and therefore no groups can conceivably benefit from special interventions to improve to the level of higher-scoring groups (Hunt and Carlson 2007:197). These assumptions are not known to be true, and much evidence suggests they likely are not. Do we not as a society disadvantage whole groups of learners if we build our policy around such misguided assumptions?

Here's another example: Adverse impact is now an aspect of discrimination law, which means that the merest indication of disparate outcomes for different groups is taken as legal proof that discrimination, which is outlawed, has occurred, and no evidence of the discrimination is required. The assumption about human nature on which this is built is the same one on which No Child Left Behind rests, that is, it is impossible that there might exist innate factors that significantly contribute to the disparate outcomes. The stakes are certainly high here, but what do we gain by pretending this cannot be so as a matter of *a priori* presumption? Shouldn't science be permitted to have a role in shaping the public understanding of this issue and thereby the rational political responses to its consequences?

Postscript on the Difficulty of Asking Some Questions about Human Nature

There is a tale worth telling in this field, one that has seemingly disappeared from the public discussion but that says much about the attitude in much of the population, and even of much of academia, on this series of scientific questions around the genomics of intelligence. About 15 years ago, a young professor of human genetics at the University of Chicago, Bruce Lahn, showed in two articles published in September 2005 in *Science* that two genes highly implicated in human brain size have undergone recent selection in our species (Mekel-Bobrov, Gilbert, et al. 2005 and Evans, Gilbert et al. 2005). One of these genes, Microcephalin (MCPH1), apparently emerged some 37,000 years ago, while the other, ASPM (abnormal spindle-like microcephaly associated), arose still more recently, less than 6,000 years ago. Both genes help determine brain size in humans. Lahn and his colleagues looked for variation across human populations in the two genes and found highly interesting results. The newly-evolved version of ASPM was most common in European and Middle Eastern populations–a category that for the authors included Iberians, Basques, Russians, North Africans, Middle Easterners, and South Asians. It was less common though still significantly present in many East Asian populations, but essentially absent in sub-Saharan African and indigenous American populations. The situation of MCPH1 was still more strikingly distinct: The derived allele was highly frequent in all populations sampled outside Africa, much rarer but still common in north Africa, and rare in sub-Saharan African populations. Lahn and his co-authors speculated that the evolutionary histories of these two genes might tell us something of utility about the development of the modern human brain, but nowhere in the two pieces did they speculate beyond the apparently recent and rapid evolution of these changes in genes related to brain development. They did not even suggest any connection to IQ and intelligence. They suggested only that the different alleles "lead to different phenotypic outcomes of the brain" (Evans, Gilbert et al. 2005:1720). That was the extent of the findings of the two studies: Suggestive of more research, but conclusive of little, and eminently careful in not overstating. This is just the way everyday science is supposed to work.

But it is well-known in current research on IQ that there is a significant correlation of brain size and IQ (Lange et al. 2010). And informally, in presentations of their research, Lahn did allude to the possible way in which these data from human evolutionary genetics might have some bearing on the effort to explain populational differences in IQ. This was enough to set off alarms in a certain element of the community of academics and, once those academic activists had begun spreading misleading interpretations of Lahn's work, in broader media publics as well. Accusations quickly emerged that Lahn was making arguments of a piece with eugenicist white supremacist perspectives from a hundred years ago, and even that he was contributing to the possibility of a modern racist eugenics. Lahn, a scholar who came to the US from China as a young adult and who seemed relatively unaware of the politically-motivated games played in much of American social science today on questions such as this one, was stunned by the response. He quickly dropped this line of research and moved into less controversial areas. In a *Wall Street Journal* interview about the move, he explicitly talked about the attacks he received for the research as part of his motivation in changing direction. That is to say, in this case it is patently clear that extra-scientific factors determined the direction of science and pressured a scientist to drop a line of research that some rejected based on purely political considerations.

A search for recent research on the microcephalin gene and its evolution in humans over the past 15 years since the Lahn controversy reveals almost nothing. In another *Nature* article from a few years after the publication of the two articles on brain size, Lahn argued for the need for "a positive attitude towards [human] diversity, including that at the group level" (Lahn and Ebenstein 2009:727), but it appears the scientific community has gotten the message on this topic from the political activists who chased him away from his earlier research program.

References

Alexander, Richard. *The Biology of Moral Systems*. New York: Routledge, 1987.

Banfield, Edward. *The Unheavenly City Revisited*. Boston: Little, Brown, and Company, 1974.

Bearman, Peter. "Exploring Genetics and Social Structure: Introduction." *American Journal of Sociology*, 114, S1 (2008):v–x.

Buss, David. *The Murderer Next Door: Why the Mind is Designed to Kill*. New York: Penguin, 2006.

Byars, S.G., Q. Q. Huang, L-A Gray, A. Bakshi, S. Ripatti et al. "Genetic Loci Associated with Coronary Artery Disease Harbor Evidence of Selection and Antagonistic Pleiotropy." *PLoS Genet*, 13, 6 (2017).

Carter, Ashley J.R. and Andrew Q. Nguyen. "Antagonistic Pleiotropy as a Widespread Mechanism for the Maintenance of Polymorphic Disease Alleles." *BMC Medical Genetics*, 12, 160 (2011).

Churchland, Patricia. *Touching a Nerve: Our Brains, Our Selves*. New York: W.W. Norton & Company, 2013.

Conley, Dalton and Jason Fletcher. *The Genome Factor: What the Social Genomics Revolution Reveals about Ourselves, our History, and the Future*. Princeton: Princeton University Press, 2017.

Dobzhansky, Theodosius. *Mankind Evolving: The Evolution of the Human Species*. New Haven: Yale University Press, 1962.

Dobzhansky, Theodosius, F.J. Ayala, G.L. Stebbins, and J.W. Valentine. *Evolution*. Freeman: San Francisco and Reading, 1977.

Esteller-Cucala, Paula, Iago Maceda, Anders D. Børglum et al. "2020 Genomic Analysis of the Natural History of Attention-defecit/Hyperactivity Disorder using Neanderthal and Ancient Homo Sapiens Samples." *Scientific Reports*, 10 (2020):8622.

Evans, Patrick D., Sandra L. Gilbert, Nitzan Mekel-Bobrov, Eric J. Vallender, Jeffrey R. Anderson, Leila M. Vaez-Azizi, Sarah A. Tishkoff, Richard R. Hudson, and Bruce T. Lahn. "Microcephalin, a Gene Regulating Brain Size, Continues to Evolve Adaptively in Humans." *Science*, New Series, 309, 5741 (Sep. 9, 2005):1717–1720

Franks, David and Jonathan Turner, eds. *Handbook of Neurosociology*. Springer Netherlands, 2013.

Gottfredson, Linda. "Mainstream Science on Intelligence: An Editorial with 52 Signatories, History, and Bibliography." *Intelligence*, 24, 1 (Jan.–Feb. 1997):13–23.

Gould, Stephen Jay. *The Mismeasure of Man*. New York: W.W. Norton & Company, 1981.

Gouriounova, Natalia and Hulbert Mansvelder. "Genes, Cells and Brain Areas of Intelligence." *Frontiers in Human Neuroscience*, 13, 44 (Feb. 2019).

Guo, Guang, Yuying Tong, and Tianji Cai. "Gene by Social Context Interactions for Number of Sexual Partners among White Male Youths: Genetics-Informed Sociology." *American Journal of Sociology*, 114, Suppl. (2008):S36–S66.

Guo, Guang, Yi Li, Hongyu Wang, Tianji Cai, and Greg J. Duncan. "Peer Influence, Genetic Propensity, and Binge Drinking: A Natural Experiment and a Replication." *American Journal of Sociology*, 121, 3 (Nov. 2015):914–954.

Haldane, J.B.S. *The Causes of Evolution*. Princeton: Princeton University Press, 1932.

Harris, Judith Rich. *The Nurture Assumption: Why Children Turn Out the Way They Do—Parents Matter Less Than You Think and Peers Matter More*. New York: The Free Press, 1998.

Hrdy, Sarah Blaffer. *Mother Nature: A History of Mothers, Infants, and Natural Selection*. New York: Pantheon Books, 1999.

Hunt, Earl and Jerry Carlson. "Considerations Relating to the Study of Group Differences in Intelligence." *Perspectives on Psychological Science*, 2, 2 (June 2007): 194–213.

Lahn, Bruce and Lanny Ebenstein. "Let's celebrate human genetic diversity." *Nature*, 461, 8 (Oct. 2009):726–728.

Lange, Nicholas, Michael P. Froimowitz, Erin D. Bigler, Janet E. Lainhart and Brain Development Cooperative Group. "Associations Between IQ, Total and Regional Brain Volumes, and Demography in a Large Normative Sample of Healthy Children and Adolescents." *Developmental Neuropsychology*, 35, 3 (2010):296–317.

Legare, Cristine, John Opfer, Justin T.A. Busch, and Andrew Shtulman. "A Field Guide for Teaching Evolution in the Social Sciences." *Evolution and Human Behavior* 39 (2018):257–268.

Leroia, Armand, Andrzej Bartkeb, et al. "What Evidence is There for the Existence of Individual Genes with Antagonistic Pleiotropic Effects?" *Mechanisms of Ageing and Development*, 126 (2005):421–429.

Levins, Richard and Richard Lewontin. *The Dialectical Biologist*. Cambridge: Harvard University Press, 1985.

Li, Yi, Hexuan Liu, and Guang Guo. "Does Marriage Moderate Genetic Effects on Delinquency and Violence?" *Journal of Marriage and Family*, 77 (Oct. 2015):1217–1233.

Mekel-Bobroy, Nitzan, Sandra L. Gilbert, Patrick D. Evans, Eric J. Vallender, Jeffrey R. Anderson, Richard R. Hudson, Sarah A. Tishkoff and Bruce T. Lahn. "Ongoing Adaptive Evolution of ASPM, a Brain Size Determinant in Homo sapiens." *Science*, New Series, 309, 5741 (Sep. 9, 2005):1720–1722.

Murray, Charles. *Human Diversity: The Biology of Gender, Race, and Class*. New York: Hachette Twelve, 2020.

Plomin, Robert. *Blueprint: How DNA Makes Us Who We Are*. Cambridge: MIT Press, 2018.

Power, R.A. and M. Pluess. "Heritability Estimates of the Big Five Personality Traits Based on Common Genetic Variants." *Translational Psychiatry*, 5 (2015):1–4.

Raine, Adrian. *The Anatomy of Violence: The Biological Roots of Crime*. New York: Pantheon, 2013.

Ritchie, Stuart. *Intelligence: All That Matters*. London: Teach Yourself, 2016.

Schroedera, Hannes, Ashot Margaryan, Marzena Szmyt, et al. "Unraveling Ancestry, Kinship, and Violence in a Late Neolithic Mass Grave." *PNAS*, 116, 22 (May 28, 2019).

Segerstråle, Ullica. *Defenders of the Truth: The Sociobiology Debate*. New York: Oxford University Press, 2001.

Stearns, Stephen. "Less Would Have Been More." *Evolution*, 56, 11 (2002):2339–2345.

—. "Are We Stalled Part Way Through a Major Evolutionary Transition from Individual to Group?" *Evolution*, 61–10, (2007):2275–2280.

Symons, Donald. *The Evolution of Human Sexuality*. New York: Oxford University Press, 1979.

Thornhill, Randy and Craig Palmer. *A Natural History of Rape: Biological Bases of Sexual Coercion*. Cambridge: MIT Press, 2000.

Trampush, J.W., M.L.Z. Yang, J. Yu et al. "GWAS Meta-analysis Reveals Novel Loci and Genetic Correlates for General Cognitive Function: A Report from the COGENT Consortium." *Molecular Psychiatry*, 22 (2017):336–345.

Turkheimer, Eric. "Three Laws of Behavior Genetics and What They Mean." *Current Directions in Psychological Science*, 9, 5, (Oct. 2000):160–164.

Wilson, E.O. *Sociobiology*. Cambridge: Harvard University Press, 1975.

—. *On Human Nature*. Cambridge: Harvard University Press, 1979.

—. *Consilience: The Unity of Knowledge*. New York: Vintage, 1998.

—. *Genesis: The Deep Origin of Societies*. New York: W. W. Norton & Company, 2019.

3 Why and How Are Humans Social?

Imagine that a group of scientists from somewhere outside our solar system arrived on Earth and observed, among the rest of life on the planet, the strange hairless primate biped that had managed to colonize just about every available ecology. What would be the first question they might ask about this animal? A good candidate would be: Why are they so consummately social?

The first approach to this question might dig in a bit more critically. Are they really so social? Just how and in what ways are they social? What precisely do we mean when we say "humans are social?" A frequent approach in mainstream American sociology is to attack "hyper-individualist American culture" by pointing out all the ways in which we rely on other human beings to live our lives. This is true enough as far as it goes. Isolated individual humans do not generally do well, the Robinson Crusoe narrative aside. But what could it mean to say that we are innately social, as such texts often do? There are a range of human societies and cultures, some of which greatly emphasize individualism of a specific kind, others of which stress the communal side of human life. In what sense are some or any of them indicative of humankind's innate sociality?

Some aspects of our lives, at least in some settings and ecologies, revolve heavily around close and frequent interaction with others, but there is much evidence too that we avoid sociality and even asocial contact with one another in many situations when we have the power to do so. We transport ourselves, for example, in most of our largest conglomerations using a means of conveyance that involves packing many of us together into relatively confined spaces, for journeys ranging from a few minutes to hours. During such collective trips, we studiously avoid talking to one another, touching one another, even looking at one other and acknowledging the presence of others. Instead, we sit quietly, in close proximity to one another, with sound producing technology in our ears to drown out the sounds our fellows might make, our eyes trained on small image-producing boxes, and we arrive on and depart from our transport devices with no greeting or farewell to our fellow passengers. During such journeys, if someone unaware of the anti-social protocol should happen to try to speak with one of the others, he will perhaps get a response, but it will probably be perfunctory and the respondent is likely to turn immediately back to his own business. If the deviant insists on talking more, the other person may ignore him, or get up and move to another location. Even if she does neither of these things, the person being addressed by the deviant will certainly recognize the deviance of what he is doing.

If we assume human sociality as given, we might be forgiven foregoing even the slightest effort to define it or investigate the origins of human social life. Frequently enough to make one wonder if there is some central script from which people are working, introductory sociology texts begin with a reference to C. Wright Mills' notion of the sociological imagination. This is the idea that the personal details of a human life are informed by the interrelationship of experience and the wider society. The notion of history is brought in as well as another force affecting individual life. But, inevitably, the context is greatly restricted and only vaguely defined. What is the wider society? The city in which one lives? Some larger social unit? What is history? What happened only in the recent past? A century ago? It is all left frustratingly vague and ill-defined, and astute readers of such texts come away suspecting there is homework here that has been left undone.

History for a Sociologist

One of the most fundamental mistakes most mainstream sociologists make is in their understanding of history and how much of it needs to be considered to have a reasonable, scientifically defensible perspective on human nature. The mainstream sociological position accepts the disciplinary separation of sociology from anthropology and essentially cedes consideration of the entirety of human history prior to industrialization to the anthropologists. This is a major error. Humans are not the way they are simply because of what has happened to us in the last 350 years. Much that is most essential in us is basically unchanged in that period. Even basic aspects of our cultures—the most plastic elements of the human world, which in principle can change fundamentally in a human lifetime or less—are astonishingly similar today to their predecessors from centuries earlier. In a few world religions with hundreds of millions or even billions of followers around the globe, basic dogmatic beliefs concerning the methods for humans to relate to the supernatural have not been altered in hundreds of years. The timeframe of analysis for most mainstream contemporary sociology is simply too narrow. Much of what built humans—that is, of what shaped our brains and their means of motivating and driving behavior—occurred while we were living in physical and sociocultural environments fundamentally different than those of the last few hundred and even of the last few thousand years. We must go back tens and hundreds of thousands of years to locate the ancestral environments in which our species spent most of its time on the planet.

A proper effort to understand the historical framework needed to effectively theorize humankind must reckon with deep evolutionary time. Sociologists perhaps cannot be expected to know much in the literature on the evolution of life generally, but at the very least they ought to be conversant with the evolutionary history of the taxonomic order Primates in which our species is situated. Some of the relevant events and dates here include:

- first primates: *c.* 55 million years ago
- last chimp–human common ancestor: *c.* 6–8 million years ago (late Miocene epoch)
- emergence of Australopithecus (precursor of the Homo genus): *c.* 4 million years ago (Pliocene epoch)
- emergence of genus Homo: *c.* 2.4 million years ago (Pleistocene epoch)
- natomically modern Homo sapiens: *c.* 300,000 years ago
- exodus of Homo sapiens from Africa into the rest of the world: 50,000–60,000 years ago
- end of hunter-gatherer era, first horticultural societies: *c.* 10,000 years ago (Holocene)
- emergence of first agricultural societies: 5,000–6,000 years ago
- emergence of industrial societies: since mid-1700s.

The typical sociological study begins at the last stage above, and typically late even in that, pretending as though the world relevant to understanding human sociality and everything else about us emerged sometime after the Second World War, or in the 1960s, or still more recently. It might reasonably be said that the very fact that there is a sub-specialization in sociology called 'Comparative Historical Sociology' is a sign of the depth of the problem. All of sociology must be comparative historical sociology or it cannot offer deep insights into what sociologists purport to study. How can we hope to know much about human behaviors and social organization if we ignore the vast majority of the history that made us, and concentrate solely on the tiny sliver of time during which we have been living in conditions utterly unlike the evolutionary conditions that produced our species?

A Primate Among Others

A look outside of the traditional social sciences is here necessary to get a fuller perspective on human sociality. The range of human societies derives, after all, from a species sample of only one. A scientific effort to properly frame human sociality would require some curiosity about how sociality looks in other species that are close evolutionary relatives of *Homo sapiens*. What do other apes and primates look like on this variable?

The primatological evidence provides a helpful framework for better understanding our complicated practices and psychological predispositions regarding sociality. There are intriguing differences in general patterns of social ties that emerge in looking comparatively at monkeys and apes. Monkey societies are typically characterized by dense matrilines, that is, females remain in their natal groups and form strong bonds with one another all through their adult lives, while males leave natal groups and maintain only weak ties to other males, females (beyond mating) and young. The structure of monkey societies is based on female relationships, and clear, strict male hierarchies inside groups also help order their social lives. By contrast, in all apes except chimps and bonobos, long-term relationships between kin are not enabled by social structure, and even in the lone exceptional ape species such relationships are limited to those between mothers and their adult male offspring. Female apes tend to leave natal groups at puberty, and males are isolated. Social ties are accordingly much weaker in apes than in monkeys.

In two of the three apes that are phylogenetically closest to us, social life is minimal. Adult orangutans spend little time interacting outside of mating activity. In eastern mountain gorillas, social life consists of groups that are seldom larger than thirty individuals, made up typically of one dominant older silverback male, one or two younger blackback males, and a group of adult females and their young. Mating in the groups is primarily the prerogative of the silverbacks. Single males, who are fairly abundant, do not interact significantly with troops, seeking occasionally to surreptitiously mate with or steal away one of the attached females. Chimpanzee social life is a bit more extensive than that of these two species. Daily chimp social life takes place in fluid subgroups drawn from larger communities that may number as many as a hundred individuals. The subgroups are however much smaller, generally numbering in single digits. Chimps are unique among the apes in that adult males form short-lived groups for marauding in conflict with other chimps, or for hunting monkeys for food, but in these male bands, hierarchy is unstable and structure is tenuous. In all these ape species, we find little long-term social structure and a general pattern in which individuals spend significant time alone or in small groups typically numbering just a few.[1]

Collective life in animal species that hunt is frequently connected to the need for cooperative group coordination during the hunting process. Yet in nearly all such species, social groups are relatively small. How is it that human societies have become so large? The best answer we have to this complex question takes the form of some external threat to the group that would advantage its size and cohesion. For humans, for much of our time on the planet one of the most significant external threats to any human society was other, neighboring societies. The record of organized warring and raiding between human social groups extends all the way back to our origins, though the historical record shows it likely increased in intensity and frequency with the rise of horticulture and the emergence of settled communities and economic surpluses. From this starting point, one can build an entire theory of the emergence and nature of morality in humans. In this view, intergroup competition directly drives intragroup competition. As soon as the former begins to grow, as a result of more small groups populating a territory in close proximity to one another, larger group size is selected for because, all other things being equal, a larger group is more able to withstand external attacks. This growth in the size of the group increases opportunities for conflict inside the group, and mechanisms that can keep this conflict in check—such as moral systems—are advantaged. Richard Alexander summarizes the situation thus:

> No other sexual organisms compete in groups as extensively, fluidly, and complexly as humans do. No other organisms at all play competitively group-against-group. Most important, so far as we know, in no other species do social groups have as their main jeopardy other social groups of the same species—therefore, the unending selective race toward greater social complexity, intelligence, and cleverness in dealing with one another.
>
> (Alexander 1987:80)

We will return to this theory of the emergence of human moral systems in the following chapter.

Illustrations of our evolutionary trajectory can be seen in the most basic elements of our physical makeup, and there are myriad aspects of our contemporary social and cultural environments made possible by those physical features that were produced by our evolutionary history. Most primates other than us live in the tropics, and *Homo sapiens* got its start there too, in the form of precursor primates living in African rainforests. Our heritage is arboreal, and adaptation for life in the trees gave us several distinctive features. Our relatively unsophisticated sense of smell and hyper-developed visual acuity, with three dimensional perspective and fine-tuned color discrimination, marks primates as an outlier among mammals. An animal the size of an ape living high above the earth in trees can feasibly do without an elaborated ability to detect predators by smell, since nearly all the predators large enough to pose a threat to such animals are earthbound. But an insufficient ability to clearly visualize, to perceive depth in the visual field as you navigate through the branches high above the hard ground, will not bode well for survival. The selection pressure of the arboreal ecology required significant rewiring of the neocortex to enable the necessary heightened vision, and this might also have indirectly contributed to the neural prerequisites for language ability. Our manual and digital dexterity, opposable thumbs, and strong and flexible wrists and shoulders were likely also all consequences of the selection pressures faced by an animal living arboreally and requiring facility at gripping and swinging on branches (Turner and Maryanski 2008). How many of us realize, as we spend hours and hours of our days working and playing with images on screens, pushing buttons and levers and keys, i.e., living in a modern society that an ultimate explanation for why such cultural inventions rather than others might have emerged has to do with the fact that our predecessors were made for life in the trees, with visual and digital acuity and skill as the means on which natural selection lit to modify the animals that eventually became us?

There is an evident intellectual requirement, then, to understand what was going on in the life conditions facing the last common ancestor humans shared with the other great apes during the late Miocene epoch, for these were the conditions that created the basic architecture of the modern human brain and thus the modern human self. Jonathan Turner and Alexandra Maryanski's *On the Origin of Societies by Natural Selection* is one of the most theoretically sophisticated efforts to place *Homo sapiens* into a fully evolutionary perspective to attempt to understand human behavioral predilections and social structures. Any serious student of this topic should read it in its entirety. What follows in the remainder of this chapter is a summary of some of its main arguments.

There is significant geological evidence that there was a major climate change event sometime around the middle of the Miocene epoch, marking a gradual but steady shift in the ecosphere inhabited by the primates who were the last common ancestors we share with chimps from warm rainforest to much cooler grasslands and savannas. Given the ecology in the earlier Miocene, it is likely that weak social ties would have been selected for in this species, as life in the arboreal cover made access to food and protection from predators relatively easy to accomplish without close social cooperation. An ecological shift to savannas meant these arboreal primates were forced to spend more time on land, vulnerable and relatively incapable individually of even detecting predators at night, with food sources significantly less plentiful and concentrated. Ape populations and species declined significantly under this ecological pressure. Early hominins faced the need of strengthening group ties to overcome the new threats from predation and the difficulty of obtaining sufficient food.

How did it happen? In fact, the logic of natural selection meant that for many species, it did not happen. The australopithecines went extinct during this period because they were incapable of effectively meeting the challenge. The emerging *Homo* genus, however, fared somewhat better. It must be remembered that this is the merciless way of natural selection. There is no mechanism forcing all individuals down the road to successful adaptation. What happens is simply that those animals least well suited to survive and reproduce in the changed environment do not do so and die without reproducing, while those better suited reproduce and, going forward, the ecology becomes more populated with other creatures that look like them. This is what happened with respect to early hominins here in the late-Miocene.

Emotions First, Reasons Later

The speculative argument as to how the *Homo* genus hominins managed to achieve stronger social ties and thereby to meet the challenges of the new ecology has to do with changes to the structure of the hominin brain. Turner and Maryanski argue that the likely route for this adaptation was through subcortical structures in the brain that govern emotional response. These parts of the brain—including the amygdala, and especially its basolateral areas devoted to pleasure, and the septum—were already substantially developed in primates. In the hominin lines that emerged from the Miocene, there was substantial growth in these structures that intensified the emotional experiences of these primates. The sensual pleasure from behavior related to the sex drive was strengthened and enhanced. This may well have been the pathway toward the emergence of the pair bonded hominin family structure, the foundational building block of a more integrated hominin social life compared to the more promiscuous and dispersed social life of most of the apes. More intense sexual pleasure that is associated with a specific individual mate is the basis of the emotional bond of love, and an institutional structure that increases the hominin ability to face serious environmental challenges emerges as a consequence.

The human nuclear family structure, which is the basis of enhanced human sociality, requires explanation precisely because this social unit is so rare in other primates. Another important question involved in the differentiation of human sociality and its possible connection to the evolution of human sexuality is the disappearance of estrus and the emergence of concealed ovulation in human females. There are a number of hypotheses concerning the evolutionary reasons why human females would have departed from the general mammalian norm of estrus, in which sexual activity is basically restricted to a short period during which the female is ovulating and external cues alert males to this situation. In several of these, the force driving the shift is the increased fitness provided to human females by mates who develop basically monogamous attachments to them and to their offspring (Barash and Lipton 2009:47–76). The disappearance of estrus in humans perhaps contributes to increased social harmony compared to some other primates in that it reduces the intensity of male competition for female sexual access, which is aggravated by the restricted time frame of estrus (Dobzhansky 1967:84).

A potential problem arises with the hypothesis that evolved capacity to experience heightened emotion in *Homo sapiens* was the catalyst for our increased social bonding. Of the four basic emotions—happiness, fear, anger, sadness—three are negative. How could it be possible, then, for enhancement of hominin brain structures controlling emotion to lead to greater social bonding, rather than provide still more fuel for aversion to social interaction? The answer is that even negative emotions have social outputs. Indeed, the role of these negative emotions in producing and reinforcing social interdependence and conformity to protective and comforting social order is a topic that deserves much more attention. Fear, anger, and sadness as primary emotions can be blended and combined in varying formulae that produce second order emotions—shame, guilt, and alienation—essential for building and maintaining social order. Shame and guilt provide mechanisms for helping enforce adherence to social norms, while alienation can call attention to a flaw in the social order and signal to others the need to act in such a way as to bring the alienated individual back into the heart of the group.

That emotional features of the human mind here get the front seat in making sense of our sociality is contrary to the basically rationalist narrative about the emergence of human social life and our most basic nature that has long predominated in the Enlightenment and social contract view so influential in the dominant modern worldview. The accepted way of understanding the place of humankind in the world, and *vis-à-vis* every other animal on the planet, in much contemporary mainstream sociology is that it is our capacity to reason, to calculate, and to subordinate emotion to our cognitive abilities that defines us and that drives our social lives. Utilitarianism of various shapes in philosophy is the intellectual backdrop of this position, and it is widely accepted as the key to understanding human nature, human interests, and the necessary structure in human social and political life in modern intellectual circles. We have the societies we have because we are basically rational creatures, pursuing our rationally calculated interests through rationalized means, and further rationalization is what will make them more perfect. Emotion has a place in this narrative, but it is clearly subordinate and frequently understood as a force standing in the way of human progress.

The truth, as Turner and Maryanski suggest, may be quite a bit more difficult and complicated. The upshot of their analysis is something we might perhaps call a new version of Durkheim's *homo duplex* argument with respect to our inclinations to sociality. "Humans," they write, "are of 'two minds' about ties; on the one hand, we crave intimacy and solidarity while, on the other, we can feel engulfed, smothered, and overwhelmed by close social relations" (2006:53). There is a fair amount of evidence to indicate just this. Our species wants not only face-to-face community and solidarity among equals, but also the freedom and individualism to escape, at least to some degree, the inevitable status rankings, hierarchy, control, and forced uniformity that come along with such community. Literary treatments of the stifling nature of overly suffocating community are often more vivid than sociological accounts, so we do well to reread Lewis's *Main Street* or Yates's *Revolutionary Road* from time to time to remind us of the ugliness and pettiness that accompanies at least some of what calls itself solidarity and community in the modern world. It may well be the case that what our species wants is neither a pure version of either one or the other of Tönnies' opposed categories of *Gemeinschaft* and *Gesellschaft*, but some composite of their most desirable features, as defined from the idiosyncratic perspective of our species at this juncture in our evolutionary history. We now find ourselves residents of contemporary macro-societies that inevitably remove us to some extent from the face-to-face community and solidarity we desire, but that in the same motion enable other things we like, i.e., individualism and the escape from the most empirically direct forms of hierarchy. A small hunting and gathering community may be the most natural social form for our species in that it best fits the various species characteristics we present as a result of our evolutionary history, but the form seems essentially extinct, and it is not clear that it is entirely realistic either to wish macro-society away or to imagine that curtailing the elements of individualism it exhibits would be an improvement on what currently exists. The ramping up of the desire for solidarity and the tamping down of rational self-interest would, minimally, require some major cultural and social engineering to implement, and there seems remarkably little mass interest in most of the West in engaging in such work. Max Weber famously recognized that *zweckrational* action has become the dominant form of social action in the West, and he saw it as mere wishful thinking to imagine that one could reverse that tidal wave by nostalgically looking backward to a time when traditional and affectual action motivations were more widespread. Even if greater solidarity could be had through a properly enlightened self-interest, how much solidarity do we want and need? This is a real and complex question, and a good answer to it requires unflinching examination of human nature in its full contours.

Many contemporary sociologists reading these last few paragraphs, given the overwhelming statistical skew of their politics in the direction of either liberalism or left progressivism, will be aghast. "So," they might ask, "it just so happens that the most fundamental truths about human nature—that our capacity for calm, dispassionate reason and objectivity is subsidiary in deep evolutionary and psychological terms to emotional investment and partiality—not to mention our basic orientation to the social world—that is, that we do not particularly crave either the selfless absorption into the collective preached by radicalisms of left and right, or the wholly self-sufficient individual of liberalisms and libertarianisms of all persuasions, but are instead tragically torn between our attraction and aversion to both, which is bred by deep mechanisms in us that mostly exceed our control—these truths are broadly consonant with a political outlook on the world reasonably described as conservative?" How can this be? Science cannot be dissuaded by the possibility that it might produce results that prove in at least some ways amenable to political ideas that intellectuals would like to avoid. If the evolutionary evidence presented here is correct, and there is no utopian solution to the complicated attitude of humans to fellowship with our fellows, then mainstream sociology will have to choose one of two options: Adjust the basic way it orients itself to social policy and advocacy to attune to the science, accepting a more measured set of ambitions for what can be accomplished at least in the short term through social engineering, or give up the rhetoric of science and accept that it is doing advocacy for an ideology, not endeavoring to understand human nature as it is.

Two short videos that recently showed up in my Twitter feed practically called out for a place in the conclusion to this chapter. In one, a human tickles a baby chimp, while the chimp playfully attempts to ward off her efforts, boasting a smile nearly indistinguishable from that of a human child. In the other, young monkeys climb up a small tree and leap off into a puddle of muddy water, then rush back up the tree to get in line behind their comrades to have another go. In watching the one, I vividly recalled playing with my own children, and when I viewed the other, I immediately thought of my own youth, my friends and I gleefully frolicking at a local pond furnished with a rope swing at the water's edge. By chance, I happened just at this same time to be rereading a book—Le Roy Ladurie's *Montaillou*—about a village in the south of France in the early 1300s, one of the first historical studies of such a distant epoch that used a variety of ethnographic record—the report drawn up by the Inquisition, which was endeavoring to eliminate the heresy of Catharism there—to sociologically document everyday life. Among the important social activities engaged in by the residents of Montaillou as described by Ladurie, frequently while discussing religion and politics, and sometimes as a precursor to romantic interludes, was the same reciprocal delousing that we can see today performed by chimps, baboons, and other primates (1979:141).

As I pondered the videos of apes and monkeys at play and Ladurie's account of the residents of Montaillou socially removing parasites from one another, I wondered what it is that frightens so many people so much about the idea that—to put the matter in bluntly evolutionary terms—they are us and we are them, insofar as we come from the same ancestor, just a blink of the eye ago in evolutionary terms. How much richer could our understanding of ourselves be if we could only build that affirmation into our every effort to plumb our own depths?

Note

1 Source material for the preceding two paragraphs is as follows: Turner and Maryanski 2008, Wilson 1975; Dixson 2012.

References

Alexander, Richard. *The Biology of Moral Systems*. New York: Routledge, 1987.

Barash, David and Judith Lipton. *How Women Got Their Curves and Other Just-So Stories: Evolutionary Enigmas*. New York: Columbia University Press, 2009.

Dixson, Alan. *Primate Sexuality: Comparative Studies of the Prosimians, Monkeys, Apes, and Humans*, 2nd ed. New York: Oxford University Press, 2012.

Dobzhansky, Theodosius. *The Biology of Ultimate Concern*. New York: New American Library, 1967.

Flannery, Kent and Joyce Marcus. *The Creation of Inequality: How our Prehistoric Ancestors Set the Stage for Monarchy, Slavery, and Empire*. Cambridge: Harvard University Press, 2012.

Kelly, Robert. *The Lifeways of Hunter-Gatherers: The Foraging Spectrum*. New York: Cambridge University Press, 2013.

Ladurie, Emmanuel Le Roy. *Montaillou: The Promised Land of Error*. Translated by Barbara Bray. New York: Vintage, 1979.

Mills, C. Wright. *The Sociological Imagination*. New York: Oxford University Press, 1959.

Tönnies, Ferdinand. *Community and Society*. Translated by Charles Loomis. London: Routledge, 1988.

Turner, Jonathan and Alexandra Maryanski. *On the Origin of Societies by Natural Selection*. Boulder: Paradigm Publishers, 2008.

Wilson, E.O. *Sociobiology*. Cambridge: Harvard University Press, 1975.

4 Morality in a Biosocial Context

One of the aspects of contemporary mainstream sociology that would most astound someone who had been present for its origins and then skipped a century or so to the present era is the relative inattention to the sociological exploration of the origins of morality. Morality was one of the central concerns of the founding generation of sociologists.

In the work of Émile Durkheim, this was especially true. His entire *oeuvre* was centered on the question of how to establish and maintain modern moral institutions capable of producing social solidarity and regulating the potential pathologies of modern individualism. He made over the length and breadth of his life's work a case for the new science called sociology, which he believed to be uniquely capable of solving the most profound moral dilemmas of modern human society. The study of morality is made difficult precisely because of its self-evident nature. Everyone assumes they know all they need to know about its workings because as moral beings they can respond appropriately to the moral questions posed in their world. But the sociologist must get to the bottom of this automatic behavior. The goal is a "science of moral facts" in which we class moral facts as we class all other "natural things" in scientific examination (Durkheim 1979:199, 202).

In his lectures on moral education, Durkheim famously insisted that the training of humans to moral systems requires a tripartite focus: The first and primary element is discipline, the second attachment to social groups, and then finally autonomous, willed valuation of the rules of morality. He was not explicit about an evolutionary perspective in this work, but what he presented fits neatly into such a framework. Morality is not an abstract philosophical system to be embraced cognitively and rationally; rather, it is something into which one is trained by superiors and the group. Children innately take to it, Durkheim suggested, because of their intrinsic natures. Like all humans, children are prone to immoderation in their desires if unchecked socially, but there are other aspects of their nature that lead them to be receptive to being commanded by elders. This is part of what he saw as their innate traditionalism, that is, a basic attraction to routine and predictability. Where might children get such characteristics? Durkheim did not speculate, but he argued that children share this complex of traits with premodern humans generally. There is "a strength in habit" (Durkheim 1961:863). An evolutionary theory might well indicate that such strength resides in the protection it has always provided, and especially so in primitive societies, against dangers and threats. Aversion to risk and adherence instead to patterns of behavior time-proven to reduce it are, all other things being equal, traits likely to confer adaptive advantage.

Throughout the twentieth century, there were some sociologists who remained focused on this topic. The sociology of religion emerged as a subspecialization that attracted scholars interested in morality. (It is a telling fact that just about every single thinker of importance in the founding generation of sociology can with accuracy be called a sociologist of religion since they nearly all took religion as a central concern). I went to a graduate program that had a faculty member and a group of grad students affiliated with the *Habits of the Heart* book and project organized around Robert Bellah, and there were others involved in this project at other campuses of the University of California system, and elsewhere. But it is patently clear today, a quarter century later, that most sociology has abandoned the intellectual effort to scientifically get to the root of the origins of morality in our species. This is not to say that sociologists have given up on moralizing. Of this, we still have much. There is an ASA section dedicated to Altruism, Morality and Social Solidarity (AMSS), but it is among the smallest of all ASA sections, with a membership lower than almost 90% of the remaining sections. More, a glance at its webpage and newsletter—and indeed its very name—indicate that the section essentially assumes we already know answers to the most basic questions about human morality, and all that remains is to find ways to expand the reach of the specific moral values of utilitarianism and altruism for which the section advocates. One looks in vain to find evidence that members of the section are at all sensitive to or aware of what evolutionary theory suggests to us about the likely origins and nature of human interests and morality.

A glance at a recent *Handbook of the Sociology of Morality* provides further confirmation of the problem. The editors of the volume note what I have just described, that is, that modern sociology has "successfully neglected [morality] for some time" (Hitlin and Vaisey 2010:3). But of thirty chapters, only one—Jonathan Turner's "Natural Selection and the Evolution of Morality in Human Societies"—takes evolutionary theory seriously and substantively. The remaining twenty-nine chapters are conventionally mainstream sociological, i.e., no biology, no natural selection, no understanding of humans as animals with animal interests. The editors claim to be interested in bringing the biological and psychological into the discussion of morality, but they neatly separate these aspects from the social part of morality that they present as the distinct realm sociologists engage with on this topic. But everything to do with the sociality of humans is an evolutionary phenomenon that is treated by primatologists and population biologists.

The topic of morality is made still more difficult for intellectuals to take up by their typical unwillingness or inability to deal with the topic objectively and make the necessary effort to leave their own moral druthers on the sidelines.

Much of the published work described in the newsletters of the AMSS is straightforwardly of this stripe. It assumes a desirable moral universe—the ideal type is perfect altruism and egalitarian rejection of nearly all status inequality and hierarchy—and then advocates for it. This is moralizing, not the science of morality. Even those who bring a biologically, neuroscientifically-informed framework to the study of morality too frequently fail to avoid this problem. Joshua Greene and Robert Sapolsky, among others, provide a useful window into some aspects of the biology of morality, and they report on a good deal of intriguing research on the relevant questions. However, their own unargued moral stances stand front and center in their discussions. Greene (2013) openly builds his theory of morality on a philosophical utilitarianism that is presented as the only intellectually defensible moral philosophical position. Sapolsky presents his own biases more anecdotally and informally, but they are no less present. He seems incapable, for example, of mentioning the work of Konrad Lorenz, one of the founders of the field of ethology and a giant in the study of animal behavior, without adding adjectives such as "odious" and lurid insinuations about the contamination of his intellectual work by his brief flirtation with fascist politics (Sapolsky 2017:83).

One does not have to argue that natural selection is moral in showing how it works, and the accurate mapping of the pathways of adaptive fitness might well become more difficult if we are so committed to a given moral view of the world that we are unable or unwilling to bracket it sufficiently. The is/ought distinction must be recognized with utter clarity here. Any view of morality that starts with the premise that the world has an innately moral logic runs into difficulty with the Darwinian evolutionary theoretical approach to morality. In the dominant theological modes of understanding morality in the West, the ultimate cause of all things is God's will, which is good, and understandable as such by humans.[1] But there is no goodness or badness evaluable in the results of natural selection and trying to find morality in nature is fruitless. Morality is a construction of humans, related to the operation of the natural world in ways that cannot be made sense of within the terms of moralities, but it is a construction that nonetheless we might very much need or at least deeply desire.

Kin Selection, Direct and Indirect Reciprocity

I sometimes use an example in class of a sociologist of the family who, in an interview in a major media publication, argued that the only way to adequately address inequality in American society at a policy level is to somehow convince Americans that they should be as invested in the well-being of all children as they are in their own. I ask my students to contemplate that and examine how realistic they think it is, what evidence there is that it is a feasible goal, and if they can think of a single parent who really believes this. (I wait until after their initial consideration of the proposal to reveal to them a detail of utmost importance admitted in the interview by the sociologist in question: She does not herself have any children, which almost certainly makes it easier for her to fail to see the difficulty the typical parent would have with her proposal). We go through the practical ways in which one would go about demonstrating not only that one adhered to this belief about moral commitment to children but also that one acted upon it—e.g., by spending at least as much on the welfare of children not one's own as on one's own children. Students are surprised, and somewhat frustrated, that they find that the belief of the sociologist rests on just about no evidence of its possible existence or plausibility on any substantive level.

How could evolutionary theory intervene in that discussion? First, it provides a reasonable explanation as to why very nearly all parents do not feel as this sociologist would like them to feel. Kin selection, as originally formulated in the work of W.D. Hamilton (1996), indicated a solution to a quandary faced by evolutionary thinkers encountering organisms that seemed to be denying a basic tenet of natural selection in the most fundamental way, that is, by evolving in such a way as to be individually incapable of producing offspring. Hamilton showed that in social insects where this is commonly found individuals who do not themselves reproduce nonetheless succeed at helping ensure that their genomes will be carried into the next generation by aiding breeding adults and their offspring who are closely genetically related to them. Mechanisms in such species prepare individuals to be highly adept at recognizing close kin and then to provide resources and protection to them. In humans, of course, the mechanisms by which we recognize kin are significantly less reliable, but they are not non-existent. Research indicates that humans are capable of recognizing kin by scent with some facility (Weisfeld et al. 2003). It also seems that deep-rooted preferences for individuals of roughly one's own physical phenotype are also a less than perfect but practically workable means by which individuals can differentiate between those likely to be more or less closely related to them (Jones 2018).

But of course human beings engage in what looks like selfless activity not only with close relatives, but also with complete strangers. This ranges from relatively low-cost behaviors like allowing someone to go ahead of you in a grocery store line or have your seat on public transport to gifting large amounts of money to a university attended almost entirely by people who are not your close relatives. Robert Trivers (2002) developed a framework for making sense of this within the purview of evolutionary theory. Kin selection can be readily seen as adaptive for individuals in that they share significant amounts of genes with close relatives. Why would we aid those distant from us? Trivers argued that in some limited number of cases it could be because direct reciprocity could be expected—that is, the specific individual you helped could at some point reasonably be expected to repay you—but what about situations in which you will likely never encounter the helped individual again, which is the case in much of our interaction in large modern societies? Trivers' argument was that several conditions that were universal in early human evolution would have made systems of reciprocity sustainable and advantageous to the actor even with those who were not direct kin. In hunter gatherer groups of perhaps 50 to 100 members, there would have been frequent occasions for interaction and repayment of services, over relatively long periods of time, in situations of significant equality with a weak hierarchy, with the likely presence of outside threats in the form of other small societies. Under such conditions, cooperation and even acts that might superficially seem wholly selfless could be frequent and mutually advantageous for those involved. As societies grow, however, direct reciprocity ceases to wield the same efficacy in producing fitness advantage for altruists. It is true that in some interactions in mass modern societies we are not completely certain how long the relationship will last and how high the likelihood that we will interact in future with the specific individual in that interaction. It could be that what we take to be a fleeting engagement—refusing to share an airport cab with another passenger because you are running late for a job interview—could turn out unexpectedly to last longer than you imagine—the man you refused to share the cab with turns out to be the employer you meet for the job interview and you do not get the job. In some such situations, it would pay to act non-selfishly, as the small cost of sharing the cab is offset by the much larger possible pay off in the job (Delton and Krasnow 2014, see below).

But Trivers argued the logic of the mechanism at work in effectively getting us to act altruistically a good deal of the time in interactions with strangers we have every reason to believe we will not see again is rather different. It is the memory of reputations on which status depends to a significant degree, bolstered by cultural and often religious narratives about morality, that come to be the overarching force encouraging such actions. Even when individuals in any given interaction are not likely to meet again—on the metro, for example—and therefore the possibility of direct reciprocity is in practical terms nil, indirect reciprocity is selected for because individuals are generally advantaged by building and maintaining a reputation as a cooperative exchange partner apparently willing to give more than they receive. Such a reputation tells others that you are trustworthy and morally sound, not a transparently selfish individual who would merit suspicion and caution, and so you and those closely related to you are likely to benefit from increased levels of beneficence from others. It is of the essence that this reputation will in nearly all cases rest on a falsehood. That is, it is given by the framework that those with high reputations in moral systems are in fact not altruists in the colloquial sense. They are exceptionally skilled at getting their own fitness advantaged through the convoluted rules of a moral system that requires the avoidance of directly selfish strategies and rewards, in a manner that can be difficult to immediately see, behavior that appears selfless, even though it is actually beneficial to the actor.

Trivers shows that deception, therefore, is a central element in systems of indirect reciprocity. People deceive one another about how eager they are to give more than they receive, and how interested they are in the welfare of others. Selection drives not only this deception, though—it also drives facility at detection of the deception. There is thus a kind of ongoing arms race between deception and deception detection in moral systems. This arms race is perhaps most visible in religion and in the mating and dating relations of men and women. Religious prophets seek audiences and followers, and the sincerity in their calling that they can demonstrate to potential cult members is the single most effective component of a successful program for recruiting them. The number of examples of such figures revealed at some point to be involved in deception is significant, and most potential followers know this. They therefore heighten and hone their skills at distinguishing false prophets from the real thing. The deceptive self-proclaimed prophet might suffer losses in their ranks as a result, and many will fail altogether to attract followers, but the high potential rewards of the occupation will ensure that many such prophets will simply work harder to maintain their advantage and stay strategically a few steps ahead of the deception detectors. Men and women in the early stages of what seems like a potentially serious relationship will know well how fraught the experience is, as mutual work is done to probe the depths of commitment, to learn about the romantic past of the lover, to expand knowledge of his or her friendship network, and to gather information from those associates as well, in short, to do due diligence in order to minimize risk of being defrauded in committing to a relationship with a cheater or an otherwise unserious partner. Here too, though, deception is so potentially profitable that it is nearly impossible to eliminate it altogether, no matter how good detection of cheating becomes.

One of the topics that can arouse torment in the mind of just about any male of our species in a long-term relationship with a woman that includes the couple's children is that of the innate difference between men and women in degree of certainty of parenthood for a given child. Any mother knows who her children are, whereas any father—unless he has had himself and his offspring genetically tested for relationship—who has not been in the company of his female companion 24/7 throughout the relationship must rely on less certain forms of knowledge to make that judgment. There is evidence that, under at least some circumstances, this opportunity for deception on the part of women—to engage in sexual relations outside the couple relationship with a male or males more genetically gifted than the long-term partner—is sufficiently enticing to draw some number of women into that deceptive behavior. As I will discuss more in the next chapter, a strong argument can be made that the dominant force behind some of the more overbearing elements of patriarchal familial orders in many societies is the desire of men to increase the amount of certainty they have regarding their paternity of the children they are helping to raise. In the world before relatively inexpensive DNA tests, there was no better method for ascertaining parentage than exercising near total control over the freedom of married women to engage in interactions with adult males other than their husbands.

What makes a particularly effective deceptive actor? If he can succeed in convincing himself that what is objectively deceptive in his own behavior is in fact virtuous and unselfish, he gets a leg up on the deceptive actor who knows, in his heart of hearts, as he is deceiving and lying about it that he is doing precisely what he is doing. The latter will tend to give himself away, unintentionally, even unconsciously, much more readily than the self-deceptive actor. So self-deception, Trivers argues, becomes a particularly effective technique to outflank deception detectors. If you are yourself convinced that you are being wholly selfless even while you are demonstrably and objectively pursuing reproductive fitness, i.e., if you believe in your own lies, you are likely to be a better deceiver of others than the merely self-aware deceiver.

The Moral Brain

Ethics and morality are ways of structuring behavior, and behavior is the work of brains, so moral systems are entwined in brain processes. Human moral life has four basic pillars: Caring, the recognition of the psychological states of others, social problem-solving, and learning social practices. Each of these pillars is rooted in basic biological imperatives driven by natural selection. We learn to care for others starting with close kin and move outward to those less closely related, and the movement outward always attenuates the degree of care in some ways. We must recognize the psychological states of others if we would attain the benefits of successfully predicting their action as part of the pursuit of our own interests. Problem-solving in a social contest is imperative to resolving conflicts concerning the distribution of scarce goods. Finally, learning social practices, through reinforcement, imitation, trial and error, conditioning, or analogy, enables us to act as members of a social group and thereby attain the various goods that accrue from such membership (Churchland 2011:9).

Care for others must originate in self-care, that is, in the instinct for self-preservation. Any organism that did not begin with a predilection for self-caring rather than self-neglect would not be around long for us to encounter (ibid.:27). The organism seeks homeostasis—which simply means an internal state within the range appropriate for comfort and survival—for itself, and in mammals concern for homeostasis of close kin also develops to a high degree. Mammalian offspring are born much more dependent than many other young. As warm-blooded animals, or homeotherms, they have the advantage of hunting during the colder temperatures at night, but this also means high energy requirements. Meeting those requirements requires a brain that is adaptive, and an adaptive brain is most effectively produced through the mechanism of learning rather than pre-programming.

Oxytocin, an ancient chain of amino acids, plays an essential role in motivating mammalian care for offspring, especially by mothers. It triggers maternal action on behalf of the infant, from feeding to ensuring they are warm and protected from harm. Endogenous opiates are also involved in maternal mammalian brains, providing pleasurable sensations when the mother suckles and cuddles with an infant. This same biological substrate that gives rise to maternal affection in mammals facilitates mate attachment, though this does not necessarily require sexual exclusivity. Subtle differences in receptors for oxytocin have been discovered to be the basis of different mating behavior, for example, in males of two species of vole. In montane voles, density of these receptors in males is considerably lower than in prairie voles. Montane males show no preference for any female partner and devote no energy to care of offspring, while in prairie voles, mating pairs stay together for life and males share parenting duties ranging from protecting the nest to caring for pups. So far as we know presently, the difference in oxytocin is the explanation (ibid.:47). And the experimental evidence for the role of oxytocin in such emotional bonding extends beyond voles. There is a fascinating research literature that shows that intranasal administration of oxytocin to human subjects *via* a spray dramatically increases their prosocial emotions and behavior, causing them to be more trusting and cooperative (ibid.:74–80).

The neurobiological equation here is of course more complex than this. The dopamine system is also an important player in prosocial behavior. But taken as a whole, these elements are a huge part of the biological substrate on which arise the emotionally satisfying sensations attached with broader social morality.

Moral Systems: Where They Come From, What Sustains Them

A full-blown theory of human morality that extends from the basics provided by evolutionary theory is perfectly plausible. In a text that is uncoincidentally virtually unknown in the social science community dedicated to exploration of this topic, the evolutionary biologist Richard Alexander (1987) has elaborated just such a theory. His approach to morality is muscular and unapologetic. He makes almost no effort to compromise with frameworks for understanding morality that reject natural selection, and he describes with crisp clarity that he intends to challenge conventional ways of approaching the study of morality in the most aggressive way. "It is a common error," he provocatively asserts, "to suppose that something additional to nepotism and reciprocity is required to account for the structure of society" (Alexander 1987:153). He cites Hume in one of the epigraphs preceding the book's first chapter to the effect that we all recognize that what we typically consider immoral behavior ("avarice, ambition…revenge") is fairly commonplace, and so anyone who told us of a society where there was none of this would not be credible. Another epigraph shows how what is framed as the unproblematically positive behavior of friendly cooperation is, with a little investigation, revealed as in the interests of vigorous competition with outgroups, who also cooperate warmly amongst themselves to outcompete us.

In virtually every system of morality, selflessness is privileged as the most moral form of action. Yet this seemingly runs against the core of the theory of human nature presented in evolutionary thinking, which is that no organisms, including humans, designed to systematically act against their own interests and in the interests of indiscriminately chosen others could hope to succeed in the competition of natural selection. Alexander unpacks the seeming problem. Our intuitions on why we act—and should, even must act—morally are not accidental. Natural selection has built them into us as a way to make it more likely we will act in just this manner. Human beings with an evolutionary theoretical understanding of their own motivations to act morally might perhaps still be able to justify and continue their moral action, but it would be due purely to a rational calculation of interests: "I do better—because my society does better in competition with other societies—by acting this way." But such a rational understanding would also have to reckon with the idea that cheating is not only possible in moral systems, but also that it can advantage individuals beyond their compliance with moral rules, so long as the vast majority of others continue to observe the rules. The mechanisms that drive moral behavior operate apart from any need for the individual to engage in such rational calculation of interests. We derive emotional benefits from acting in ways prescribed as moral, most directly in the form of positive reinforcement from others, first and most importantly from the others from whom our young selves most desire approval, that is, parents and other close relatives. It should also be recognized that in all of human history, we find a more or less universal belief in supernatural entities that tend to be defined in such a way that they can be expected to reward moral behavior with gifts of great value—perhaps immortality, or in earlier religions, the gratitude and recognition of the spirits of the community's deceased ancestors.

Alexander acknowledges that the study of human social activity by reference to evolutionary biology has in the past been a charged endeavor, with some critics attacking it vigorously and some practitioners inadequately separating their scientific analyses of human behavior from ideological goals. But over time, he suggests, these controversies of the birth of sociobiological interest in the human world will fade and the term "sociobiology" with all the controversy surrounding it will disappear, while the work of studying humans from an evolutionary biological framework will go on. One of the advances we have made over earlier evolutionary understandings of human action has to do with the current rejection of the simple contrast between the biological and the social or cultural, though many conventional social scientists continue to deploy this outdated framework. Biology includes psychological and cultural aspects of behavior. Minds are biological systems constructed by natural selection. Culture is also a product of the imperatives of natural selection. Another advance concerns the fact that we know more now about how to effectively adhere to the scientific imperative to move beyond proximate/" how?" questions to ultimate/"why?" questions in examining human behavior. Translating and reducing proximate analyses of human behavior to the ultimate level—sometimes in the form of reductionism—is fundamental to scientific understanding. Although this focus on the proximate is most typically found among non-biologists, even some in the biological sciences fall into this way of thinking about human behavior—Alexander criticizes Stephen Jay Gould and Richard Lewontin for failing to avoid this error.

Moral and ethical questions exist only because of conflicts of interest. The relationship of the interests of separate individuals, whether conflictual or congruent, is "the warp and woof of the fabric of society" (ibid.:34). If there were no such conflicts, we would not need moral systems. Moral systems exist to aid in the resolution of these conflicts. What are human interests? Sociology has spent considerable time and energy on this question, but mostly unfruitfully. Alexander rightly starts with scarcity. Not all can get all their interests realized. There are finite possibilities to realize interests, and so some will be realized and some will not. How are interests determined? First, they are at root individual, not collective. Collectivities can, in a fleeting, always fluid way, operate as apparently motivated by something we might call group interest, but that is always subject to collapse and transformation based on the shifting of individual interests inside the group—e.g., a change of leadership, or the dissolution of a coalition. Much conventional sociology—especially but not only of a Marxian persuasion—posits a false consciousness version of interests in which subjects do not know what their real interests are and are deluded as to what they take for their interests. Alexander says that people are usually pursuing their real interests, but they do not typically know precisely what those interests are, and, as we have already seen in the work of Trivers, there is in many cases an evolutionary advantage to organisms like us that can effectively fool not only others but also themselves about their interests.

But surely there must be some actions that meet the vernacular definition of altruism, which is to act in such a way as to simultaneously aid the interests of others and neglect or even harm one's own interests? Alexander allows that such behaviors may exist, but only as errors, destined to work ultimately against an individual's interests and therefore having no chance of being selected for in any ecology we can easily imagine. Joseph Lopreato (1984) suggests an alternative typology for social behavior with respect to rules of moral systems that endeavors to make specific theoretical space for such acts. All behaviors, with rare exceptions, colloquially described as altruistic would for Lopreato be classed into one of the following four categories: Three sub-varieties of favoritism–nepotistic, reciprocal, and phenotypical–and ascetic altruism. The first three are behaviors that involve aiding, respectively, close biological kin, those bound by formal or informal rule to reciprocity, and those with phenotypical features homologous to the actor. Those three are all covered under Alexander's theory. Only ascetic altruism is behavior that disadvantages the actor at the expense of the aided individual. Lopreato does not say how common he imagines this fourth category would be. It is in the behavior of some religious retreatants and mystics, and in for example the tiny percentage of the population that deliberately avoids parenthood and all significant contribution to the welfare of relatives, that we would look to find evidence for this type. Though Lopreato does not see such acts as Alexander does, that is, as resulting from mistakes or defects inherent in the actor, he does agree with him that such forms of behavior must be rare, and the vast bulk of human behavior is contained in his first three categories.

The baseline needed for an understanding of moral systems, and particularly for the practice of what we colloquially call altruism, is the way the means of resolving conflicts of interest shifts in humans as a result of the fact that our chief threat comes from groups external to our own group. As Alexander puts it, "rudimentary moral systems (indirect reciprocity) will appear where outside threats most powerfully dictate group cohesion, when such threats are combated best by complex social organization within the groups, and when the actions of single individuals or small subgroups can threaten, from within, either the group as a whole or its most powerful elements" (Alexander 1987:80). As noted in the previous chapter, intergroup conflict drives intragroup conflict, and as the former increases, the advantage of larger size in groups grows, and as groups grow in size, opportunities for interindividual conflict rise, necessitating systematic mechanisms for resolving those conflicts, preferably in the interest of group cohesiveness in competition with other groups. Interests are still individual and reproductive here, that is, nothing has changed fundamentally in terms of how we conceptualize human interests. But the group dynamic is what gives rise to moral systems that will ensure that individual conflicts of interest are worked out in ways consistent with the basic rule of natural selection—individuals pursue inclusive fitness—and the cohesiveness of the social group at the same time.

The central task of moral systems, then, is to regulate intergroup competition in a way that does not damage the group overall *vis-à-vis* other groups. Indirect reciprocity is the mechanism Alexander argues for as the solution to this problem. He schematizes social action in such a way as to systematically trace its motivations and consequences in a way that would be testable empirically. There are four possible categories of social actor. The first is the brazen cheater or open egoist, who rejects the moral system's rules and nakedly pursues self-interest. These people are typically avoided by others if possible, and they are thought of badly in general. Ultimately, this mode of action is probably the worst at achieving the end desired. Self-interested ends are harder to realize if others see you as an undesirable exchange partner and your reputation is damaged as a result. The sneaky cheater/insincere altruist also consciously tries to achieve his own self-interest but attempts to hide this fact from others. They can sometimes succeed at this deception, but when they are discovered, the game is up, and others are avid to detect such deception and reveal it to others because such acts bring good will from others. The honorable egoist pursues the rules of the moral system—indirect reciprocity—but does so with the conscious understanding that he is nonetheless pursuing his own interest, and he refrains from conceiving of himself or his acts as selfless and altruistic. The self-professed altruist not only acts according to the rules of the moral system but also internalizes the notion that doing so requires selflessness and perceives himself to be acting out of a concern for others and not for self. Self-deception is typically involved here.

Alexander's argument enables a new and insightful view of the role of religion and its evolution in human history. The invention of gods is, in his analysis, a way to facilitate the interests of members of one social group against one or more other groups. Early religions are always and everywhere tribal religions, in which the god represents the tribe and is indifferent to or actively opposed to the gods of other peoples. Universalist religions likely emerge as the religions of subjugated peoples in a majority society dominated by ethnic or racial others. They operate effectively in the interests of such groups, as they challenge dominant groups to acquiesce to the moral and ethical claims of the downtrodden to be treated as equals. But if those religions achieve cultural triumph—as, for example, Christianity did throughout Europe by the early Middle Ages—and become the faith also of the dominant segments of societies, it is not at all clear that the force of selection can operate in the same way as it did earlier, as now an ethic of universal brotherhood cannot but work against the higher status of those strata. Given this dynamic, we might well surmise from Alexander's approach to religion an evolutionary inevitability of the relative decline of universalist religions in class-stratified societies once they are taken up by dominant groups, barring other developments or specific allowances in the religions involved for the maintenance of social systems of hierarchy, e.g., as models of the hierarchy in the heavens.

The game of morality, then, according to the logic of Alexander's argument, is to pursue one's individual interest by means of the cultural mechanism of indirect reciprocity. Individual humans have been selected for many tens of thousands of years, at least, to follow this blueprint, and the way in which natural selection works with respect to moral systems is typically the same way it works with respect to other fundamentally essential aspects of behavior. That is, if an individual can be made to act "altruistically" through subconscious decision-making, this is more efficient and effective than relying on conscious thought. What we call the moral conscience is in fact the subconscious aspect of our rule learning with respect to moral systems.

Indirect reciprocity makes no sense in a world without reputations. Direct reciprocity will necessarily become more rare as a society grows in size, as many of the people you meet on a given day and might act toward with at least mild altruism—allowing them to enter the intersection first or to merge into traffic from a ramp—will never be met with again. Except in rare situations, you will not be harmed by cheating in any one given interaction. In the scenario already given of declining to share a taxi only to later learn the individual rejected turns out to have a significant possible favor that he now will not return, we typically do not forego the short-term cost for the long-term gain, precisely because the odds that we will meet him again in such a situation of possible exchange are too long, and we are not programmed to think in those terms. We see the advantage to be gained at present—avoid the discomfort of making small talk with a stranger, ensure that you will not be delayed by any rerouting based on his destination—and take it (Delton and Krasnow 2014).

Only social reputation can produce the "selfless" behavior—here, in the eyes of the cabdriver and any other onlookers. As already described, a morally positive reputation makes people more likely to trust and desire to interact and exchange with you, which means you are more likely to benefit materially, and this also aids your efforts to materially advance the interests of your children. Your reputation can even carry over to a significant degree to reputation for your children—"she's from a good, morally upstanding family, and so we run little risk in deciding in her favor"—and this provides that same benefit of exchange networks to them. There is evidence that both men and women rate generosity and selflessness highly in evaluating potential mates, though there are sex differences. Even if there are no onlookers, the individual can gain reputationally by telling the story to others who were not present at the act. This includes telling the story to himself, mentally, a way to better train the self into a pattern of behavior that seems to conform to the moral system's imperative while ultimately aiding his own interests. Alexander liked to tell a story that illustrated both the desire most of us have to avoid admitting that our moral behavior is motivated by what he argued drives it and the difficulty of escaping its logic in moral systems of indirect reciprocity:

> Theodore H. Hubbell, a friend and colleague [of Alexander's] at the UMMZ, was troubled by the contention that human behavior is by and large self-interested. Once he thought he had a good counter-example, and approached Alexander: "Dick, this morning on my way to work I saw a caterpillar moving across the sidewalk. I knew it had fallen off its host plant, so I picked it up and replaced it on the proper plant species. Was that not an act of pure altruism?" To this Alexander replied, "It may have been, until you told me about it."
>
> (Summers and Crespi 2013:138).

Alexander closes his argument with the hope that, even given the rigor of his case for the origins of moral behavior, we might somehow manage to achieve the global cooperation he believed was necessary to end the central human political threat at the time of the book's writing, that is, the Cold War and its danger of global destruction through nuclear war. But his hope was dashed at the same time that his theoretical framework was fully vindicated in the way that problem was resolved. For it was not through the emergence of a universalist and rationalist selflessness, but through the competition of interests, and the victory of a culture that successfully made itself stronger through a more solidary moral system than its opponent that the Cold War was concluded without mutual destruction in the rain of atomic bombs. The very human nature built by natural selection to be essentially impervious to selfless action managed, this time at least, to accomplish an end beneficial finally to all involved.

Note

1 I must leave to the side the fascinating question of whether this is the only way to see Darwinian logic from within Judeo-Christian moral understandings. Briefly, I believe it is not. Dobzhansky wrote a remarkable book (1967) on the compatibility of the neo-Darwinian view and a version of Christian morality, albeit one that presents the question of God's ultimate moral plan for the universe as more inscrutable to humans than is commonly encountered among the religious faithful. But this is a topic for an entire book of its own.

References

Alexander, Richard. *The Biology of Moral Systems*. New York: Routledge, 1987.

Bellah, Robert, Richard Madsen et al. *Habits of the Heart: Individualism and Commitment in American Life*. Berkeley: University of California Press, 1996.

Churchland, Patricia. *Braintrust: What Neuroscience Tells Us About Morality*. Princeton: Princeton University Press, 2011.

Delton, Andrew and Max Krasnow. "An Independent Replication that the Evolution of Direct Reciprocity Under Uncertainty Explains One-shot Cooperation: Commentary on Zefferman." *Evolution and Human Behavior*, 35 (2014):547–549.

Dobzhansky,Theodosius. *The Biology of Ultimate Concern*. New York: The New American Library, 1967.

Durkheim, Émile. "On the Learning of Discipline." Pp. 860–865 in Parsons, Talcott and Edward Shils et al., eds, *Theories of Society*. Glencoe, IL: Free Press of Glencoe, 1961.

—. "Introduction to La Morale." In Pickering, W.S.F., ed. *Durkheim: Essays on Morals and Education*. London: Routledge & Kegan Paul, 1979.

Greene, Joshua. *Moral Tribes: Emotion, Reason, and the Gap Between Us and Them*. New York: Penguin, 2013.

Hamilton, W.D. *Narrow Roads of Gene Land, volume 1: Evolution of Social Behaviour*. Oxford: Oxford University Press, 1996.

Hitlin, Steven and Stephen Vaisey, eds. *Handbook of the Sociology of Morality*. London: Springer, 2010.

Jones, Doug. "Kin Selection and Ethnic Group Selection." *Evolution and Human Behavior*, 39 (2018):9–18.

Lopreato, Joseph. *Human Nature and Biocultural Evolution*. Boston: Allen & Unwin, 1984.

Sapolsky, Robert. *Behave: The Biology of Humans at our Best and Worst*. New York: Penguin, 2017.

Summers, Kyle and Bernard Crespi, eds. *Human Social Evolution: The Foundational Works of Richard D. Alexander*. New York: Oxford University Press, 2013.

Trivers, Robert. *Natural Selection and Social Theory*. New York: Oxford University Press, 2002.

—. *The Folly of Fools: The Logic of Deceit and Self-Deception in Human Life*. New York: Basic Books, 2014.

Weisfeld, Glenn E., Tiffany Czilli, Krista A. Phillips, James A. Gall, and Cary M. Lichtman. "Possible Olfaction-based Mechanisms in Human Kin Recognition and Inbreeding Avoidance." *Journal of Experimental Child Psychology*, 85 (2003):279–295.

Part 2

Basic Categories of Human Differentiation

The second part of this book considers in detail a range of topical areas in sociology from the biosocial perspective outlined in the preceding chapters. Here, I will lay out some of the concrete ways in which the theoretical approach outlined in the first part of the book provides real insights and knowledge about human behavior and social structure that cannot be found with more traditional, biophobic perspectives.

A good deal of mainstream sociology can be accurately summarized as a series of statements, entirely or nearly so from the viewpoint of various critical theories, about a triumvirate of topics: Race, Class, Gender. Innumerable courses and publications in the discipline deal with one or another of these topics and their role in producing social identities and social inequalities. The three ASA sections that explicitly name one, another, or all three of these as their topic of concentration are among the sections with the highest memberships.

One of the things that unites most professional sociologists is an agreement that all three of these topics are examples of wholly socially constructed categories that illegitimately distinguish and divide individuals and put them into different positions in morally unjustifiable hierarchies. These categories are understood as having no real basis in objective reality—they are simply ways in which groups of people decide to group people together and distinguish them from one another. The moral fervor with which these beliefs are held is intense, and opposition to ideas about these three topics that depart in any significant way from orthodoxies is equally vigorous.

I therefore want to be clear in advance of the chapters to follow about what I am doing and not doing in them. I argue in this second part of the book that each of these conventional sociological topics can be fruitfully examined from the perspective of the evolutionary theory I have described in the first part of the book. I criticize the conventional view that the topics examined in the next several chapters are pure social products of the particular social environments found in modern human societies and cultures—though I certainly do not at all deny that sociocultural forces play a significant role in shaping them—and present evidence that each of them reflects in significant ways sociobiological facts about our species that complicate the conventional view considerably.

What I do not do in the chapters to follow is make, or even suggest, any arguments about how societies ought or ought not to rank people according to these or other categories. That is not a scientific question, although it is true that a scientific position on the nature of those categories might be seen as a useful tool to help inform the discussion about social policy related to them. There are no political statements about race, gender, or class in this book. I do not advocate for any way of thinking about social and political policy with respect to those topics, and I suggest nothing at all about what is the most moral position to have on them. Absolutely nothing I say here can legitimately be interpreted as support for any crude statements about the moral ranking of any human identities or groups as "better" or "worse" or "superior" or "inferior" or anything of the like. It troubles me that our current discourse on these topics has gone so completely off the rails of reasonable and careful argument that I feel compelled to say such obvious things explicitly, but I am so compelled and so I am saying them.

I add one final thing, that should not have to be said given a reader with a basic knowledge of how the logic of population distributions work, but that is so important that I want there to be no possible misunderstanding. Nothing that I have to say in the following chapters, or indeed anywhere in the book, about social groups—whether sex or gender groups, class groups, racial and ethnic groups, cultural groups, or any others—and the ways in which the theory of natural selection might help us think about their origins and natures has anything necessary at all to do with any particular individuals who might be in one or another of those social groups. Individuals are unique, and they are found throughout the wide distribution of any groups in which we classify them, and knowledge about the distribution alone cannot tell us anything about specific individuals. Indeed, we do not even yet know the distributions of different groups on given genetic variables, so we have as yet no ability to scientifically talk about e.g., how the variances for groups will affect propensities of individuals in groups to be concentrated in one or another part of the distribution. Though it is likely that distributions of different groups on some variables may differ significantly, there will be substantial overlap between distributions even in such cases, and any random individual in any group might be literally anywhere in the distribution.

The final chapter of the book is deliberately reserved for the topic—culture—that was for most of my early career the lodestone of my approach to sociology. When I was considering graduate study in sociology, I was rather schizophrenically interested in both culturalist and evolutionary approaches and applied to two programs that represented in my young mind the best of those two ways of doing sociology. I was accepted at both and made the hard decision to choose one, which launched me on a career trajectory that was for ten years or so quite consistent with the culturalist graduate training I received there. However, and despite a thoroughly biophobic graduate training, I never really cured my schizophrenia during graduate school or afterward, and I continued reading in and thinking about the kind of sociology represented by the program I had not chosen even as I published in the sociology of my graduate training. Over the last half of my nearly two decades as a professional sociologist, it has become more and more clear to me that the evolutionary/biosocial program is the way forward, and that cultural sociology, like the other established mainstream approaches to the discipline, leads ultimately to a scientific dead end. The book therefore concludes with an account of the reasoning by which sociobiology triumphs scientifically over cultural sociology and a critique of a critique of sociobiology by one of the leading defenders of the culturalist approach, the anthropologist Marshall Sahlins.

5 The Sex Difference in *Homo sapiens*

In his 1973 book *The Inevitability of Patriarchy*, the sociologist Steven Goldberg writes: "Sex is the single most decisive determinant of personal identity; it is the first thing we notice about another person and the last thing we forget" (1973:229). The first claim is bolstered by mountains of evidence, and the truth of the second can be tested and affirmed by every reader just by thinking of people he or she knows. Goldberg's book is a rigorous defense of the reality of a meaningful sex difference in our species with significant consequences for just about every facet of our behavior and social structure. Today, not yet 50 years from Goldberg's date of publication, it is impossible to imagine a book like his emerging to anything but massive assault by the scholarly establishment in sociology, and indeed the state of accepted pedagogy in his discipline on the topic he addressed is now nearly the precise opposite of the central argument of his book.

How did we get this far away from something this basic?

Basics of Sex Difference in *Homo sapiens*

An answer to that question requires backing up first to briefly outline the basics of the sex difference, in humans and more broadly in many other animals that are not far away from us on the chain of life. What is it, at root, to be male or to be female in the human species? At its core, the difference is biological, though environmental factors—including social institutions and cultural practices—play a role in how the difference manifests itself at any particular time and place. Human beings, like all primates, all mammals, and indeed nearly all vertebrates, reproduce sexually, which means it takes sex cells from both a male and a female to propagate offspring. We get the term "sex" from an etymological root that means "section or division." The process by which life on Earth, which originally reproduced entirely asexually, evolved sexual reproduction is a fascinating and difficult question. J. Maynard Smith's (1978) systematic effort to explain how and why this would have emerged is bolstered by mathematical density but its basic theoretical logic is elegantly simple. Genetic recombination from the genomes of two parents would have provided a decided advantage in avoiding harmful mutations and giving possible protection against ecological threats from invasive organisms. The sex difference is at its core a response to the nearly universal competition in the world to stay one step ahead of opponents. It is unsurprising then that some competition is inherent in relations between the sexes as well, though there is also substantial cooperation. It turns out thus that proponents of radical feminism are correct about one thing at least: Competition is omnipresent in the natural world. But what their efforts to understand the relations between the sexes in our species do not show is an understanding that sexual reproduction is a mode of response to competition that involves both competition and cooperation, and each in abundance.

The first point that needs to be made clear on the sex binary is that it is indeed a binary, and only exceptionally rare individuals can be reasonably understood as occupying some position between the two poles of male and female. This fundamental fact has been made murkier in recent years by varieties of feminist thought with an interest in challenging core facts of science for ideological reasons. The relatively new academic solar system of women's studies is the primary place where one encounters the attack on sex as a binary, but it has come to be important in sociology and some other social sciences as well. The claim is rooted in the writing of Anne Fausto-Sterling, a radical biologist whose work she admits has been thoroughly shaped by what she has called her "1960s street-activist heart" (2000:19). She suggested there are actually five sexes in *Homo sapiens*: Males, females, "true hermaphrodites" with one testis and one ovary, male pseudohermaphrodites with testes and "some aspects of female genitalia" but no ovaries, and female pseudohermaphrodites with ovaries and "some aspects of male genitalia" but no testes (1993:21). She guesses that some significant percentage of the human population—a few in every hundred—might be in one of the three "intersex" categories, and indeed postulates that there is an "infinitely malleable continuum" between them.

Few other biologists share this view. It was quickly pointed out in the wake of Fausto-Sterling's original five sexes essay that she had been deceptive in her estimate of the frequency of intersexuality. She counted, for example, phenomena such as Klinefelter's Syndrome (biological males with an extra X chromosome, or 47, XXY), Turner's Syndrome (biological females with only one X chromosome, or 45, X), and a number of other conditions typically not recognized as intersex. Sax (2002) points out the chicanery at work in Fausto-Sterling's effort to get the number of the intersex to the figures she claims. She directly discusses only cases that are clearly examples of intersex according to her categories, but in her statistical estimate she includes five much more frequent phenomena that are not among her three intersex categories. One of them alone, late-onset congenital adrenal hyperplasia (LOCAH), which involves the overproduction of adrenal androgens, accounts for nearly 90% of Fausto-Sterling's figure of 1.7% of the population that is intersex. But LOCAH is not an intersex phenomenon. Many individuals who have it are never diagnosed because the symptoms are so mild, and all who have this condition are born with typical male or female genitalia that correspond to the male and female genotypes. Nearly all such individuals go through puberty with the typical sexual development for their genotype, as the condition does not manifest in women typically until the early 20s and in men much later than that.

That Fausto-Sterling makes such intellectually suspect moves to boost her total is representative of the spirit of intellectual inquiry in radical circles on such issues. That she is often the only biologist cited in women's studies and sociology of gender textbooks says much for what can be expected in those sources in the way of scientificity. The true estimate, per Sax, of intersex in our species is roughly 0.018%, about 2 in every 10,000 births, or about 100 times lower than Fausto-Sterling's estimate. Another way to put this is that more than 99.98% of humans are clearly either male or female in terms of biological sex.

Sex is determined not merely by the presence or absence of a Y chromosome in the twenty-third pair of chromosomes, as was previously believed, but by the presence or absence of a particular gene typically but not always found on that chromosome, which is known as the SRY (Sex-determining Region Y) gene. This gene, which can occasionally be transferred to the X chromosome during recombination resulting in an XX individual with male sex characteristics, produces a protein that prevents what is otherwise the default development by the fetus of female reproductive organs and initiates the development of male gonads. Testosterone is key to the development of male sexual characteristics. Artificial introduction of sufficient amounts of this hormone in even young children produces post-puberty male sex characteristics. In a two-year-old who showed signs of entering puberty—an enlarged penis, the growth of pubic hair, development of facial acne—it was discovered that the child's father was a bodybuilder who was using a testosterone cream on his arms and shoulders to build mass, and in carrying the child in his arms, he was transferring enough testosterone to the child to produce these effects (Freeman and Heron 2007:962).

In many animals with a sex difference, including humans, eggs and pregnancies are significantly more expensive than ejaculates. A human spermatozoon is a tiny entity, the head perhaps 5 micrometers long and 3 micrometers wide, with a length of about 50 micrometers, nine-tenths of which is tail. A human oocyte, by contrast, is enormous, a circular cell perhaps 100 micrometers in width and several thousand times larger in volume than the sperm cell. It is in fact the largest kind of cell in the human body. In purely material terms, then, the individual egg is much more resource-intensive to produce. The sperm is considerably less onerous to make, and males make many magnitudes more sperm than females do eggs. Females of our species stop producing eggs before they are born, and they shed most of the eggs they are born with before they reach puberty. Ultimately, only a few hundred eggs will be ovulated by the typical woman during her lifetime. By contrast, males continue to make new sperm throughout life, though they suffer a decrease in motility and an increase in mutations that can make fertilization problematic as a male ages. The quantity of sperm that can be released in a single ejaculation ranges from several hundred million to over a billion. During a man's lifetime, he can produce half a trillion spermatozoa.

The sperm cell is so economically constructed because it is basically just a DNA delivery device, while the egg, once fertilized, has much more work to do to get the embryo to parturition. The simple fact that the gametes are so materially unequal in value, and that females of our species carry embryos to term, and once pregnant face a closing down of all reproductive strategies other than those related to the embryo they are currently carrying contributes to the inevitability of the emergence of different reproductive strategies for males and females. Mothers typically make a larger parental investment—which simply means that they spend more energy and time in making and caring for offspring—than fathers. Parental investment affects fitness, and because investment is typically different for the two sexes, fitness as well tends to work itself out differently in males and females. Focusing a great deal of energy and time on any single offspring inevitably means less energy and time can be invested in others.

An example from another species close to us in evolutionary terms is illustrative of the difference. In orangutans, adult males and females spend almost no time near one another when they are not mating. A male and female might spend a few days in the same space in the course of a reproductive episode, during which they will engage in multiple copulations. They then separate permanently. The comparative calculus of parental investment is quite different for the two. The female will carry the fetus for a period of 8 months, during which time she is its sole source of nutrition. It will weigh one-fortieth of her body weight at birth. The infant will remain with the mother for perhaps as long as nine years before it goes off to live its own solitary adult life. The male orangutan's investment of time and energy, by contrast, beyond the few days he spent exclusively with the female during courtship, consists of the minute amount of semen to produce the pregnancy, which was replaced within hours after the act. More than 90% of mammals have a sex differential on parental investment that is a lot like this—great investment for females, comparatively quite little for males. In such a scenario, the female's potential reproductive success is likely to be small, and the central factor that determines how effectively she will realize reproductive success is the number of pregnancies she can carry to term, not the number of male mates she can procure. For males, the contrary is true: potential reproductive success is relatively large, and the determining factor on his actual success is the number of partners with which he succeeds in procreating, not the number of sperm he can produce. The Bateman-Trivers principle nicely summarizes this typical scenario: access to the sex that makes the greater parental investment—females—tends to become the crucial resource limiting the fitness of individuals of the less investing sex—males. Selection thus favors competition among the latter for access to mates, and this means males pursue females and compete with one another for sexual access to them. Sexual selection is thus typically more important in the evolution of males than in the evolution of females (Freeman and Heron 2007:404).

The reproductive decision-making in most of historical time for our species for females, once pregnant, fundamentally revolves around these options:

Dedicate resources to the existing embryo and later child for the full nine months, several years of nursing, and then until the child is itself reproductively viable, or, if at any point in this trajectory the child is perceived to be unlikely to reach the point at which it will be reproductively viable itself, abandon it and then pursue another such opportunity. Other direct reproductive efforts will be fully eliminated for at least the first several years if she decides to invest in this child. For males, the reproductive decision-making options are different. They can, after ejaculating the sperm that fertilizes the egg, choose to invest more resources in helping the embryo thus produced to reach reproductive viability, though even if they do not do so there is significant likelihood that the mother will so invest, and he can also pursue other reproductive opportunities with other women at any point during the pregnancy and nursing of that offspring. The number of such opportunities will be limited by many factors, but they will frequently be greater than the zero presented to the female.

What Else Follows from the Logic of the Sex Difference in Evolutionary Terms?

The consequences for our species of these early differences in the basics of sexual reproduction are substantial. A sexual division of labor appears in every historical society about which we know anything substantive in which females are overwhelmingly the early caretakers of children, as they carry in their bodies the only food the child can safely eat in the first period of its life, at least until the development of infant formula in the mid-nineteenth century, and hormonal factors that are well-documented—oxytocin plays an essential role—produce the emotional bases of mother–infant bonding. Males in our species do spend much more time in the labor of child care and nurturance than any of our close primate relatives, but in every human society the disparity on this between men and women is significant, and it is directly explicable through the way in which reproductive interest works out in a situation of anisogamy. In comparative analysis, men simply stand potentially to gain more in fitness by investing relatively less in each offspring than women.

The gender feminists who are the mainstream of the view of gender and sex difference in contemporary sociology argue that parenting asymmetries and other aspects of division of labor that run along the line of the sex difference are a pure product of social structures, socialization, and cultural values. But again this is a tremendously common mammalian framework. In the vast majority of all mammalian species, it is females that provide most of the parental care, while males typically provide only relatively little, and often none. Among those primates closest to us, orangutans are the most polar on this dimension, as males abandon females abruptly after mating and have nothing whatever to do with the offspring, while the mother cares for it for nearly a decade afterward. Human males have historically been much more invested in terms of providing resources for their offspring, and the evidence suggests this was so even in early periods of human societal evolution, but the general pattern is recognizable in us too (ibid.:404).

The sexual division of labor extends beyond child nurturance to other aspects of work in all societies. In the form of human society we believe to be the earliest, the hunter-gatherers, the basic structure of labor is neatly binary and mapped directly on the sex difference. Men do nearly all the hunting, and women do nearly all the gathering. The ecological conditions presented to the hunter in sub-Saharan Africa during the Pleistocene present an obvious explanatory framework for this development. Any early society that experimented with gender egalitarianism in the harsh, exhausting labor of hunting—which in this world looked nothing like hunting today, and involved long, dogged pursuit of prey in hot sun over rugged terrain with the goal of tiring the animal out sufficiently over time to get close enough to dispatch it—would have faced unsurmountable problems with repopulation. The young women most fit to hunt would have been pregnant or nursing for a good deal of their potential hunting years, and they would have risked their own lives and those of their unborn and infant children if they had participated in hunting with any frequency. In a society at subsistence with a birth rate of bare replacement, it would have been a demographic catastrophe to have a significant number of otherwise viable pregnancies lost by mothers exposed to the harsh conditions of hunting (Kelly 2013:218–224).

Much of the rest of the background of contemporary occupational disparities is likely to have some relationship to these facts of sex difference. Take, for example, the disparity between the number of men and women involved in political occupations in contemporary American society. Feminist analysis asserts that this has to do at bottom with domination or oppression of women. But what if it is simply the case that, generally speaking, men are more likely than women to be interested in running for office? Evidence suggests substantial differences in ambition to run for political office on the part of men and women equally matched in demographic terms (Fox, Lawless and Feeley 2001; Lawless and Fox 2005). Why would such differences exist? The suggestion of feminist scholars is that it is all social and cultural influence and if you just changed this, the ambition gap would disappear. But in the contemporary societies that have, by most measures, the greatest degree of legal and social gender equality, we do not see what should happen if this view were right. For example, in Scandinavian countries that are by most measures the most gender equal societies on earth at present, fewer women pursue STEM degrees than in some third world countries (Stoet and Geary 2018). Evolutionary theory suggests that when you increase the freedom of men and women to make choices about careers and such in an environment in which—through a generous welfare state—you have reduced material necessity, you tend to see more clearly the differential preferences of the two sexes, and on average men and women tend to have distinctly different preferences about what academic subjects and career fields they find interesting. As I discuss below, there is evidence in male/female differences in brain structure and consequent distribution of personality characteristics among the sexes that go a significant way toward explaining this.

Asymmetries not only in parenting roles, but also in mate preferences emerge from the basics of anisogamy. Understanding how physical attractiveness is perceived by the two sexes requires considering several kinds of criteria. Certainly, some cultural differences in how attractiveness is perceived and rated exist, but they should not be overestimated. There is a great deal of consistency across cultures in how males and females evaluate attractiveness in members of the opposite sex. Absolute or universal characteristics are present in all human societies. These consist of clear complexion, cleanliness, clear eyes, sound teeth, firm muscle tone, and luxuriant hair (Symons 1979). It is clear why these characteristics would be universally preferred in the natural selection perspective—all are basic signals of health, and a healthy mate is a good predictor of healthy offspring. Both sexes show great interest in partners who are seen as kind and unselfish, but men rate physical attractiveness significantly higher than women in importance in mates while women are typically more concerned than men about a mate's status in material terms, i.e., his level of social prestige and economic well-being. There is an asymmetry too regarding preference for height in mates: men prefer mates shorter than themselves, and women prefer men who are taller than they are, and the preferred height differential is significantly higher for women. All these asymmetries can be made sense of in evolutionary terms. In deep evolutionary time, pregnant or nursing females would have been quite vulnerable and in need of resources that would have been difficult for them to supply alone, and a male partner with an investment in the safety and well-being of mate and children and the resources and/or physical attributes necessary for contributing to same—height is a reasonably good proxy for formidability—would have been an effective method of meeting this need.

The game of sexual interaction between men and women divides the players into two categories: Those who seek sexual favors (men) and those who grant them (women), typically in exchange for other resources. In a slight modification of a common saying, we might well say "Man proposes, Woman Disposes." This is not of course to say that women are indifferent to the pleasures of sex. It is simply to recognize that the sex difference presents a situation in which they have more to lose in any possible sexual interaction than men do, at least in a world in which there is no reliable birth control, which describes the world we lived in for most of our time as a species, and so the default of the system is to make them relatively more conservative and defensive about access to their reproductive resources. In early human societies, and among chimpanzees as well, meat is a frequent exchange resource males offer in trade for sexual access. When we evaluate human societies as to the structure of their gift-giving around sexuality, the skew is clear: The vast majority of societies are those in which only men give gifts—flowers, candy, jewels, etc.—to accompany courtship. Some much smaller number of societies show evidence of male and female gift-giving in courtship, but men's gifts are typically of significantly greater relative value. There are no human societies in which women's gifts are of greater value than men's or in which only women give courtship gifts (Symons 1979:257–258). Nowhere is the skew more evident than in the realm of prostitution and sex for money. In no human society known to history has it been the case that women gave money to men to induce them to have sex with them with anything near the frequency of the reverse practice. Pornography too bears the evidence of the skew. Even pornographic media specifically invented in the heady days of sexual liberalization of the '60s and '70s—*Playgirl*, for example—for use by heterosexual women ended up attracting a customer base too inconsequential to keep the media in business.

Sexual attractiveness in our species works across the sex difference in a skewed way and there is evidence of a great skew in terms of the percentage of each sex that is viewed as a potential sexual partner by members of the opposite sex. Both men and women can be arranged as groups in bell curves according to their level of sexual attractiveness to the opposite sex, but the two distributions look rather different. Symons (ibid.:268) gives us a visual representation of what this would look like. The median of the female distribution (as seen by men) is much higher in desirability than the median of male attractiveness for females, and the female distribution is much narrower and higher, that is, the average woman is seen as quite desirable and the range on either side is more restricted than for men. Women tend to see a large percentage of men as of very low sexual attractiveness, and the male sexual attractiveness distribution is much broader, with a comparatively much smaller number of individuals seen as highly desirable compared to the distribution of female attractiveness. The typical male views virtually all women, with the exception only of the most decrepit and physically deformed, as potential sexual partners from whom they would require no compensation for the sex, while the typical woman views only a small minority of men as potential sexual partners not requiring a counter-gift for the sexual transaction. This sets up an exchange imbalance that is a structural fact of all human mating systems. There is intrasexual competition in both sexes—that is, both men and women compete with members of their own sex for sexual access to the most desirable partners of the opposite sex—but virtually all females are assured of finding some sexual partners, even if these options are not their optimal ones. By contrast, some significant number of men are doomed to a situation in which they lack the resources—physical attractiveness, personality characteristics, high occupational status and income—to attract even the least desirable sexual partners. The bottom line for this observation in terms of how sex pairing works is that "the average male considers that he is being granted a favor if he copulates with most females, whereas the average female considers that she is granting a favor if she copulates with most males." Most men in this framework are individuals "with whom no female considers it a favor to copulate" (ibid.).

The size dimorphism in our species—the fact that men on average are significantly larger and physically stronger than women—which correlates with so much else in the way of sex asymmetry is theoretically addressable *via* the different mode of operation of sexual selection. The extent of the dimorphism is not often fully understood by those who have not looked at real data—it is in fact the case that 99.9% of females fall below the median for males on upper body and total muscle mass and strength (Lassek and Gaulin 2009). One must be careful to assume too much, of course, for a considerable amount of the size dimorphism in our species could well be a product of what has been passed down to us through ancestors. It is even possible that the size difference across the sexes has decreased in our species compared to earlier primates from which we descended—the fact that other close primate relatives such as gorillas and orangutans have much larger such size differences than we do suggests just such a scenario is plausible. However, there is also considerable evidence to suggest that male-on-male competition for reproductive opportunities with females, working itself out not only in direct male-on-male fighting and bluff based on size, but also on female choice of larger males who by their size give indication of greater ability to control resources for offspring as well as superior ability to protect a female's offspring against any aggression from others, played a role in the fact that men are typically larger than women. Again, the Bateman-Trivers principle elucidates theoretically why this should be so. Males, the sex that typically invests less in parenting, pursue females, the sex that typically invests more, and fight with each other over sexual access to them. The size dimorphism and the dynamic just discussed of some significant quantity of males unequipped with the resources needed to attract willing mates also contributes to the unfortunate phenomenon of male-on-female sexual violence, on which more shortly.

We must be careful to avoid assuming that the concept of sexual selection makes it possible to make rigid, unchanging statements about male and female strategies. In our species, there can be little doubt that male aggression and propensity to violence are higher, which correlates with other findings and suggests that sexual selection has been operating much more intensely on males, but this does not mean it is completely absent in females, nor does it mean that the Bateman-Trivers principle allows us to monolithically talk only of males as sexually promiscuous and of females as wholly uninterested in promiscuity. There are strong reasons to see our species as the result of both these aspects of sexual selection previously detailed, but there is also plenty of reason to suspect that male investment in a given spouse and their offspring may in many situations have been at least as advantageous as the pursuit of many matings, and likewise female promiscuity as a potentially successful reproductive strategy can be seen with some frequency in our species. As Sarah Blaffer Hrdy notes, the old cliché "Higamous, hogamous, woman's monogamous, Hogamous, higamous, men polygamous," which contains a powerful kernel of truth in evolutionary history terms, must nonetheless be qualified. In ecologies wherein no good male providers/protectors can be found or when resources are generally unpredictable, and when male political power over women is not sufficiently strong to prevent it, female reproductive fitness will be advantaged by promiscuity (Hrdy 1999:233). This is all complicated in contemporary analysis by the fact that we no longer live in the natural environments that produced much of our sex difference-based behavioral characteristics, so evaluating the contemporary fitness effects of given behaviors becomes a difficult task. The contribution of such a set of theoretical frameworks to the scientific task of explaining why contemporary people do what they do, however, remains quite substantial. Our minds are, after all, still fundamentally the minds built by this evolutionary history, however much the environment has changed, and it is not at all impossible to envision a social science that could fruitfully set about the task of evaluating the interaction of this Pleistocene mind with a post-industrial ecology.

The Sex–Gender Relationship

Much of the mainstream sociological effort to understand what sociobiologists refer to as the sex difference hinges on the rhetorical move away from the category "sex" and toward the category "gender." The way this move works is that sex is considered a purely biological category and therefore abandoned to the biologists—though as we will see, there are sociological efforts to take on sex by undermining most of the aspects of the term that come to us from biological science—and gender is defined as a purely social and cultural category that has no concrete, determinable relationship to sex. But arguing for a neat separation between sex and gender is denying much empirical evidence present in real human beings and real human societies. The social and cultural categories we produce to talk about gender difference are unavoidably intertwined with the biological basis of the sex difference. It is not completely arbitrary that, for example, in virtually all societies, assertiveness is more frequently a character trait assigned to the masculine gender than to the feminine, and there are good reasons to believe that individuals at one end of the sex binary are much more likely to be more immediately and powerfully drawn to the gender role that includes assertiveness as one of its core traits. There is an evolutionary logic that describes why this would be the case. Anisogamy and sexual selection alone tells us that we might expect this. Social and cultural forces can of course shape that in many ways, and even work against the logic of anisogamy and sexual selection. But to completely separate the gender categories from the sex categories, as some contemporary feminist-based writing recommends, represents a fundamental misstep if the goal is the most accurate scientific knowledge of why human behavior and social arrangements look as they do.

The use of the term "gender" to refer to the social role differences that emerge from sex difference is relatively recent. Cultural anthropologists did early work to show the category's variability across cultures, but some exaggerated a few exotic examples into a theoretical effort to undo the binary nature of the gender axis. The alignment of the sex binary with gender is a cultural universal, with some variation at the edges. Just as male and female are the two categories of the sex binary, with the vast majority of nearly 99.99% of empirical examples occupying one of those two positions, so with gender, albeit with more room for variation in the latter. Some cultures here and there will have an additional gender category or two, occupied by a small number of individuals, perhaps permanently, perhaps as a transitional or temporary identity, between the two main categories, but the two main categories are an overwhelming cultural universal.

Robert Trivers (2011:163–4) describes the "extraordinary verbal one-step" that has taken place in recent decades to effectively replace sex with gender as the basic term for denoting the sex binary in our species. Both terms have a long history of several centuries in English of being used to describe the male/female difference, but gender has an even longer association with linguistics and grammar. Languages frequently assign gender to nouns, in ways that are extremely variable and can appear arbitrary—e.g., "moon" is feminine in French and Spanish, masculine in German. Trivers suggests that the more relativist and amorphous of the two terms has recently come into more widespread use even in the social sciences, at the expense of exactness, precisely because of the political desires of those advocating for its use. 'Gender' implies that maleness and femaleness are less clearly separated than is implied by "sex."

A remarkable tendency in the gender-focused literature is to find a singular piece of a culture that emphasizes some degree of gender fluidity or mixing of binary categories and then endlessly discuss that, without recognizing that in everyday life the society or societies at issue are remarkably consistent with the binary sex/gender norm in our species. It is certainly true that specific cultural means of displaying maleness or femaleness can differ significantly. For example, one society might award status to males who excel in contests of physical prowess, while others highly value males who demonstrate the ability to calmly solve logical and technical problems, especially as they involve important technologies. But in both societies, it is a virtual certainty that men will be, on average, larger, stronger, more assertive, and more aggressive, with predictable general consequences for social relations and social order. Universally observed higher levels of male aggression will work themselves out in behavioral predilections slightly differently in this culture compared to that one, but the fact that males are universally more aggressive across cultures remains. To fixate on the specifics of minor, limited cultural variation as a contribution to the effort to distract attention from the universals in human nature is indicative of an anti-scientific motivation in this field.

Contemporary cultural anthropologists and scholars influenced by their conception of culture have greatly developed and refined the skill of emphasizing similarities in the sexes, or differences that seem to value women over men, while ignoring much more obvious differences that help produce more or less traditional differences between the sexes in social position and status. Sometimes the acrobatic rhetorical moves they must make to do so are evident in their own words. One scholar writing about Nahua society in pre-Columbian Mexico points out that Nahua women who died in childbirth were assigned high status, and economic life among the Nahua involved both men and women. This is evidence then, he goes on, that Nahua notions of gender elude the typical gender/sex mapping we see in the evolutionary perspective. But elsewhere in his discussion he admits, in an aside, that men were uniformly in control of the two central institutions in Nahua society, politics and religion, and that only men could be warriors, the single highest status position available (Sigal 2011:14).

Sometimes, the move to relativize among radical gender/sex scholars involves imputing the inadequacy of the entire intellectual culture from which evolutionary theory emerges. In a book that is a favorite of advocates who would reject the elements of the category of sex I have discussed here more or less altogether, one learns that "[t]he fundamental problem is that our academic disciplines are all rooted in Western culture, which discriminates against diversity" (Roughgarden 2009:3). Gender as a category is in some of this work remade to the point that it does not even do any of the work that would be needed to serve as a substitute for sex. Some feminist sociologists see the entire business of gender categories as little more than "a larger process whereby groups and categories of people create and perpetuate distinctions defining themselves and identifying 'others.' Boundaries serve to create and maintain inequalities in many spheres" (Epstein 2006:45). That there are substantive, statistical differences between the two genders, in all societies, that are essentially those of the two sex categories "male" and "female" is explicitly rejected by some of these thinkers. It is claimed that "there is considerable debate as to whether [males and females] have different basic emotional and cognitive attributes [that lead them to] act and feel different with regard to a wide range of behaviors and characteristics." If there are differences in behaviors and predilections, then they must be produced by socialization and culture, not biological sex difference. There are, we are told, "obvious and subtle controls that assign females and males to social roles…rules (either set by law or tradition) are supported by values that specify such ideas as 'Women's place is in the home taking care of children,' or 'Real men will fight to protect their family and country'" (ibid.:46).

Let us use just these values to evaluate this argument and illustrate its weaknesses. There is no question but that cultures and socialization do indeed make such statements about values—that is, that it is good that women be interested in taking care of children and that it is good that men be interested in protecting their kin and their countrymen. Many known human societies embrace just these two value claims and socialize their members accordingly. Why is it the case though that we so frequently find these values assigned in this way to the genders and not, say, similar values with the genders reversed? Where are the societies that advocate that men should be the primary caretakers of children and women should take the central part in physically defending kin? One can certainly conceive what such cultures might look like, but why are there no such societies in existence? To be sure, there are societies that say both men and women should have some interest in both duties, but the skew is empirically just as the author notes, and yet somehow she does not have the curiosity to ask why there are literally no counter-examples with the gender values reversed in all of human history.

The typical feminist answer to that question is something along these lines: Men control power in most if not all societies, and they prefer doing the fighting to doing the child-rearing. But why would that be so? If it is true that this is just a game of men as a class looking out for their interests against the interests of women as a class, why would they choose the more lethal of the two activities? Every year in every single society around the planet, it is safe to say that many more young men lose their lives fighting to protect their families and their countries than women lose their lives taking care of children. (This balance shifts if one builds into the latter number the quantity of women who die in childbirth annually, but that is related to a fact of biology that could not be addressed by the choices of men regarding social roles.) War is more prestigious than taking care of children, says the feminist, and that is because it was men who invented the status systems. But it seems that men would be still cleverer and more manipulative orchestrators of power if they could only devise a way to turn child-rearing into the most noble and prestigious of activities, and relegate war to something of less importance, and then force women to go and get themselves killed in the latter, while they watch the kids and get all the prestige. Why would they deliberately choose the more dangerous activity as their own, if their patriarchal power were presumably sufficient that they could have done as they pleased?

We also must note that there is a lot of empirical evidence to show that women and men do have typical preference schemes about these two activities, and they line up with the cultural values our feminist sociologist notes. That is, women on average are much more interested in caring for small children than men are, and men on average are much more interested in physically fighting to defend those close to them than women are. Further, as I will discuss shortly, the evidence shows that this propensity to gender differences on such matters is there early on, even before socialization and culture have had time to do much work to shape these values. What could be causing that, since it clearly cannot be the gender role socialization that only comes later?

The great advantage of evolutionary theory is that it has a framework for understanding this state of affairs, whereas gender feminist theory has next to nothing to offer here, beyond meta-claims about the gender bias of the science showing the sex differences. Females of mammalian species like us are more interested in and more likely to bond with infants precisely because they give birth to them, and their bodies have been modified by natural selection to enhance the likelihood that they will have significant interest in caring for them, and in most cases significantly greater interest in that care than biological fathers do. The gamete difference previously described is only the base of this explanation. Once you have the differential interest in the welfare of offspring produced by that, you will tend to get neurochemical and hormonal forces pushing in the direction of care and bonding. The difference is not categorical across the sexes, but it is there as a sufficiently statistically obvious fact that it makes sense to distinguish the sexes on this variable. Male and female responses to infant cries, for example, illustrate both the similarities and the differences. When the tenor of the infant's cry is of distress—as in "I am under serious threat"—both mothers and fathers respond quickly. But if the infant cry is not a distress signal but just an attention-getter—"I want someone to hold me now"—mothers and fathers do not respond equally. The former are much more likely to respond quickly, and even to be unable to ignore the cries without anxiety (Hrdy 1999:212).

As I was working on this chapter, I had a real world experience that gets at this point in a concrete way. I was sitting in an auto dealership while the exhaust system on my car was getting some much needed attention. There were only men in the room—perhaps ten of us total—until a young woman entered, accompanied by an older woman, and holding a baby carrier. For half an hour, the two sat there in the room with the baby, surrounded by men, none of whom ventured even to look into the carrier to see the baby, much less to inquire about the child of its mother and, presumably, its grandmother. Then two more women, also apparently an adult mother–daughter pair, came in and went to the counter to tell the clerk they wanted to have their car inspected. They sat down on the side of the large room opposite the other pair of women with the baby. In well under a minute after sitting down, though, they saw the baby carrier. Immediately, the younger of the two women got up and walked briskly over to the other two women and began to lavishly compliment the child, cooing, oohing and ahhing, avidly remarking on the baby's attractiveness. The older of the two recently-arrived women joined her quickly and chimed in with the baby talk and admiration. A few of the men in the room looked over once or twice noncommittally; most seemed not to have noticed this interaction at all and kept busy at whatever they were already doing. None of the men made any comments about what was going on among the women in the room—beyond, of course, those in the form of the hastily typed notes on the goings-on that were recorded by a certain male social scientist who happened to be in the room.

All of us will likely, with a little effort, be able to recall similar experiences. Try to imagine, though, a human culture in which the sex reactions were reversed. Theoretically, such a thing should be possible if the reactions seen in this kind of scene are completely caused by socialization and cultural norms. Yet imagining such a thing as a culture in which men ooh and ahh over babies while women sit at a disinterested distance is the stuff of comedy skits, not human societies we can imagine as plausible.

The Male Brain, The Female Brain

There is significant scientific documentation of the physiological differences in the typical male and female human brains, as well as experimental evidence indicating the differences in the typical range of behavioral and personality characteristics that correlate with these biological differences. Though there is still much to learn, we also already have a fair amount of information that indicates that some significant contribution to the behavioral and personality differences is made by the physiological differences. Evolutionary theory provides compelling models for understanding how and why such differences would have emerged in our past.

The reader might fruitfully be reminded here that all that is about to be said on this topic has to do with what we know about the distribution of attributes for the two groups "human males" and "human females." None of what follows can tell us anything necessary about specific individual men and women, or about what is possible or impossible for any given individual man and woman to achieve. Statistical differences between two groups A and B do not imply that some members of group B cannot be closer than some members of group A to the median of group A if the two groups' distributions overlap significantly—and, again, this is the case in the male and female distributions for most of the psychological and cognitive traits on which we might want to analyze and compare men and women. The point here is simply that there are distributions for both sexes on a range of physiological and behavioral features, and, while there is some overlap, the male and female distributions are not the same.

In a nutshell, research on sex differences in brain structure indicates that female brains tend to be more geared to producing empathy, while male brains tend to be more geared for systemizing. This means, very broadly, that the typical female is more interested in and adept at understanding and interacting with other people than the typical male, while the typical male is more drawn to and capable of understanding and working with systems of various kinds than the typical female. Empathy is not just the ability to understand affect or emotional content, although it is that as well. Someone who effectively empathizes can engage also in the cognitive work of deciphering the intentions and the motivations of others to accurately understand situations and predict scenarios. It is a cognitive skill in addition to an affective skill. In the appendix to his book on this topic, Simon Baron-Cohen (2003:186–199) gives an example of a well-known visual test that illustrates empathizing skill. It shows a number of human faces, or more accurately parts of human faces in various expressive moods and gives a set of descriptive emotional options from which to choose, e.g., surprised, anxious, aroused. The images show only the eyes of the individuals, so the entire emotional affect must be read from just that one cue. Females typically score significantly more accurately on such tests than males. The evidence of greater empathy in females is consistent even in those so young as to have been only minimally exposed to any gender socialization in this direction. For example, infant girls as young as twelve months respond more avidly to the stress of others than boys the same age do (ibid.:33). Likewise, the systematizing propensities of boys are evident quite early. Baron-Cohen reports on a study done in his lab on facial interest patterns in one-day old infants in a hospital in Cambridge. The newborns were given two objects to look at—a woman's face or a mobile that did not look at all human, dubbed "the Alien" by the research team. The infant boys were significantly more interested in looking at the mobile, while the infant girls preferred the woman's face. One-year-old boys have been found to be more interested in watching a video of cars than one with talking human heads with the sound turned off, while one- year-old girls have the opposite preference (ibid.:84–85).

The study of male and female human brains for structural evidence of these behavioral and psychological differences is proceeding rapidly. Among other facts that are well-corroborated already are the greater volume of the male brain in regions related to decision-making and greater connectivity of the female brain in regions related to sociality. The evidence on male and female cognitive ability is complex and indicates that males typically exceed females in some such abilities, while females exceed males in others. In mathematical skill, for example, females score higher on computation and verbal problem-solving, while males score higher in abstract math and problem-solving. The evidence to this point of male and female brain size difference shows similar variation in typical size of a number of different cortical regions. Our knowledge of precisely how each of these regions contributes to cognitive skills is still developing, but this is doubtless a field that will advance a great deal in the coming decades. Another key finding of sex difference in the human brain is the confirmation of a broader theoretical framework known as the Greater Male Variability Hypothesis. In a range of cognitive skills, males and females have roughly equal median scores, but males tend to spread out more at both extremes. In overall math ability, for example, the average for males and females is similar, but there are more males than females who are exceptionally good and exceptionally poor at math. Brain physiology conforms to this pattern as well (Ritchie, Cox, Shen et al. 2018).

There have been attempts to challenge these findings. One that was avidly taken up in some circles in academia and in the popular science press extended from an article that made a claim for what it termed "the mosaic brain" (Joel, Berman, Tavor et al. 2015) The argument in this study, which had a data set of 1400 male and female brain MRIs, was that male and female brains only very rarely map at one end or the other of a continuum of features typically assigned as male and female. The reporting from some media sources on the article was that it had definitively put an end to the arguments about significant group differences in male and female brains, and by extension to the argument about a biological component to male–female differences in behavior and personality. But the methodology in the piece has subsequently been called into question by other researchers. One critical response (Del Giudice, Lippa, Putts et al. 2015) demonstrates that the article's method for classing data on brain structures discards most of the information useful for recognizing differences, using only a small number of carefully selected variables from among many available in the data to examine sex differences. The piece also defines what is required for an individual brain to qualify as "typically male" or "typically female" too strictly and concomitantly exaggerates small levels of variability, claiming they do not permit group classifications that they clearly do allow. By the classificatory logic of "the mosaic brain," a hypothetical male individual who, given a range of possible leisure activities coded for sex-typical preference, showed no interest in scrapbooking or cosmetics (both coded "female") and was interested in boxing, construction, and golf (all coded "male") but ambivalent about video games (also coded "male") would have been classified as showing "substantial variability" on the male–female continuum. Most tellingly of all, Del Guidice and his co-authors went on to apply the classificatory schema used in the mosaic brain article to a database of facial morphologies of three monkey species of different genera, that is, a group of animals characterized by anatomical differences much larger than the intraspecies gender differences in *Homo sapiens*. The article's method was able to accurately classify into discrete species categories only an exceedingly small percentage of the monkey facial data, which pointed to a "lack of validity" in this method (ibid.:3). At several levels, it was clear that the effort in the mosaic brain article was to construct a method for distinguishing male and female that would make it possible to discount clear and meaningful evidence of categorical differences. The methodological sleight of hand of the piece was not remarked on in the breathless reception it got among many scholars and journalists eager to deny this aspect of the sex difference.

Violence and Aggression

Perhaps the behavioral and psychological predilection that has been most studied on which there is a clear sex difference that cannot be explained by recourse to differential gender socialization is aggression and violence. We have data on this dating from the 1920s that have gotten more fine-tuned over time, in a range of methodological styles, all clearly illustrating a greater male propensity for serious aggressivity. The most nuanced contemporary research shows females more prone to indirect aggressivity—i.e., emotional manipulation, social ostracism, gossip—while males engage in both verbal and physical direct aggression at significantly greater levels. The pattern is clear: The more extreme the form of violence and the more likely it is to elicit a violent response, the greater the male exclusivity in the behavior (see e.g., Archer 1994; Kortüm and Heinze 2013). In the United States, about 90% of murders are committed by men. Men are also overwhelmingly likely to be the victims of murders and other serious violence. Around 75% of murder victims are male, with roughly 90% of those male victims murdered by other men. Women killing other women accounts for only about 2% of all murders annually. More, there is a clear evolutionary framework for understanding what motivates a good deal of serious male violence. Male rates of killing start rising at around age 15 and reach their highest levels between the ages of 20 and 29. The rates remain high until the late 30s, when they begin to decline steadily. The period of highest propensity to violence is precisely the time when males enter the years of peak reproductive competition. The typical scenario of many of these acts of male-on-male violence is fairly obviously connected to intrasexual selection and male status hierarches related to mating. A fair amount of these cases consist of e.g., two males directly fighting about sexual access to a young woman or about perceived acts of "disrespect"' that threaten to decrease the status of one or the other of the involved males (Daly and Wilson 1988).

Despite the state of the evidence on this, there are still challenges to these facts from mainstream feminist sociology. In a typical example of such pleading (Epstein in Chafetz 2006:53), it is claimed that the voluminous research on sex difference and aggression fails to show a distinct pattern by citing one article (Frodi, Macaulay and Thome 1977) that purportedly claims "it is difficult to make a clear assessment of differences." But in fact, the article in question, in its first sentence, takes the following very conventional position: "Clear sex-related differences in the volume of physical aggression exist in the real world, with men as the pacesetters…and most often the victims" (ibid.:634). We know a good deal about what causes this difference. Testosterone is a key ingredient, but it is not alone. Males are typically higher in testosterone than females, which helps explain the sex difference in aggression, but what about trying to explain the differential levels of aggression and violence within the male population? As described in chapter two, it turns out that testosterone levels act in concert with the levels of various stress hormones to produce higher propensities to serious aggression. Cortisol is especially important in tandem with testosterone. High testosterone coupled with low levels of cortisol create the likelihood of high levels of aggression, whereas high testosterone and high cortisol tend to be characteristic of men who are "courageous but not reckless" (Churchland 2013:145). And it is important to recognize that cortisol levels are not solely about genetic factors, but also have to do with the effect of the social environment. A still more fine-tuned analysis can be attained by adding in serotonin. A high testosterone, low cortisol male with low serotonin tends to make for impulsive aggressivity, i.e., wildness, whereas the same male with high serotonin is more likely to produce planned aggressivity. This last profile is that of many psychopaths.

The greater propensity of males to violence extends to sexual violence. This of course is one of the most charged topics in the public sphere today. The great majority of men in modern societies will face significant structural barriers and forceful socialized inhibitions that will prevent them, in any scenario, from participating in acts of sexual violence. If however the potential costs of sexual violence are low—e.g., there is lax criminal prosecution of demonstrable sexual assaults, or the dominant cultural framework gives a fair amount of license to male sexual aggression—we can expect the incidence of male-on-female sexual assault to rise. Even in societies like those in the modern West where both criminal law and culture consider rape and rapists as meriting vigorous opposition and prosecution, the natural selection framework suggests that at least some portion of those males at the bottom end of the scale of desirability are likely, in an environment that presents opportunity, to consider this immoral and criminal mode of action. What kind of environment presents opportunity? An example might be one in which many young and desirable females are present in public settings frequented by many lower status males who are seen as undesirable by most females—say, a college campus situated in an area adjacent to neighborhoods where such lower status males live and work.

Few topics better demonstrate how gender feminism has warped the public understanding of a serious matter than this one. Traditionally, this was understood according to a logic that informally at least conforms well to evolutionary theoretical knowledge about the sex difference. The short summary of the conventional view runs thus: Men are the perpetrators of sexual assault, and women are its victims. When men are victimized, it is by other men, typically in situations—e.g., prison—where women have been systematically removed from the interactive field more or less completely. This observation conforms to the basic details of sex difference outlined above. Rape is a phenomenon that exists in many species beyond ours, and the factors involved are basically the same everywhere. If sexual opportunities can be increased by forcing oneself on a smaller female who is unlikely to be able to successfully resist or inflict serious damage on an attacker, and there are no larger social forces—law, cultural codes that oppose aggressive male sexuality, or vigilant male relatives of sexually desirable females—to prevent or avenge acts of male sexual aggression, at least some males will likely pursue such opportunities. The higher the cost of resistance, the less likely females are to resist. The higher the cost of forcing copulations, the less likely males are to employ force. In our species, these behaviors are understood unquestionably as acts of violence, and they are criminal acts to be condemned and judged accordingly. Evolutionary theory tells us that they are also sexual acts—non-consensual ones on the part of the victim, to be sure, but clearly motivated by aspects of reproductive fitness on the part of the attacker.

Feminists deny this. The classic gender feminist argument here (see e.g., Susan Brownmiller 1975) is that rape is, in purest terms, an act of collective violence by all men against all women, designed to terrorize the class of women into subordination to the class of men. The barest of examination of this claim shows its indefensibility. If the act is designed merely to terrorize women, why does it involve genitalia at all? Why would men not simply physically attack and beat women in a systematic way, as, say, young male chimps do to young female chimps? Note well that in the chimp example, the explanatory logic that best makes sense of why this would happen is that it conditions females to be submissive and to avoid resistance when males endeavor to achieve sexual access during the female estrus, so the violence is ultimately about sex. But in the feminist case about human rape, it would seem far more rational, if the goal is simply political domination of women as a class, for men to engage in the most readily available form of violence that offers the least likelihood that they too will be harmed in the interaction. After all, how much easier is it for a vigorously resisting woman to harm a male attacker if his sensitive genitals are exposed than if they are not?

The feminist definition of rape arises in part because the victim's perspective—from her perspective, the act is obviously an act of violence with no content that could meaningfully be called sexual beyond the fact that it involves genitalia—is taken to be the only one useful for defining the situation. If our task is moral judgment, that might well be the correct stance. But if we are looking to produce scientific explanation, we are required to explore the situation more carefully and fully. And this is especially so if we hope our science can contribute to meaningful and effective policymaking. Rapists demonstrate in their own testimonies mixed motives much of the time, but sex is nearly always a part of the mix (Symons 1979:282–283). Fortunately, everyday life presents few low risk opportunities for such assaults, and therefore their frequency is almost certainly much lower in modern societies than it has been in our deeper history. However, higher risk situations in modern societies do exist, and they include those—e.g., college fraternity parties—in which heavy use of alcohol is involved and sexual titillation in the form of close physical proximity and touching is present.

There has in recent years been something of an effort on the part of scholars and advocates influenced by the mainstream feminist sociological view of relations between the sexes to focus more on male sexual victimization (Stemple and Meyer 2014). Rape—that is, unwanted sexual penetration—of males certainly does occur, of course, but when it does, it is almost always other men who are the perpetrators. The actual crime statistics are stark in terms of the sex difference in sexual violence: Female victims of male perpetrators are far and away the largest category. Bureau of Justice statistics suggest the female rape rate is five to six times higher than the male rate. All data indicate that the overwhelming majority of individuals who commit rape with female or male victims are males. Perpetrators of every kind of sexual violence against women are male at a rate significantly above 90% (Smith, Chen, Basile et al. 2017, table 3.4). The great majority of rapes of male victims happen in jail or prison. While Bureau of Justice statistics suggest the rape rate for men in the American population is around 0.3 per 1,000, or 0.03%, it is as much as 4% or one in every 25 for men in jail and prison. It is sometimes claimed in feminist academic circles that much higher numbers of men have been victims of criminal sexual violence, but they do not report because they fear stigma. This speculation that significant numbers of victims of such crimes do not report would seem hard to believe on its face, at least if we accept that such sexual victimization of men is even nearly as traumatizing as it is described for female victims. One would have to believe that the stigma the male victims face for reporting is greater than the physical and psychological trauma of the crime itself.

Another contemporary strategy for finding significant numbers of male victims of sexual assault is to posit that females frequently sexually assault males. This requires the creation of a new and broadly-interpreted category of such offense. Some surveys now include a category of sexual violence called "made to penetrate." This term indicates

> times when the victim was made to, or there was an attempt to make them, sexually penetrate someone without the victim's consent because the victim was physically forced (such as being pinned or held down, or by

> the use of violence) or threatened with physical harm, or when the victim was drunk, high, drugged, or passed out and unable to consent.
>
> (ibid.: 19)

But the category has many confusing aspects. How can someone be made to penetrate someone else against his will if he is "passed out"? The functional difference in male and female genitalia inevitably plays in a role in the difficulty of simply translating a definition of sexual violence that works in one case (a woman is passed out and a man sexually penetrates her) to another case (a man is passed out and a woman forces him to sexually penetrate her). To say the least, this is something that is likely not easily achieved. When one looks closely at the varieties of such acts reported by some researchers, one finds, for example, that the vast majority (in the case referred to here, fully 70%) of the alleged male victims of made to penetrate victimizations were participants in acts classed as "alcohol/drug-facilitated" and those acts classed as involving any kind of force are a small minority of such cases (ibid.: 18, 26, tables 3.1 and 3.5). In other words, even when you create a new and amorphous category designed specifically to facilitate finding male victims of female sexual aggression, the overwhelming majority of the acts that can be fitted into the new category are not defined by the purported victims as acts of force or violence. How many of these purported acts—reported, it merits noting, in a phone survey, one of the methods of social research most notoriously prone to validity problems in which respondents embellish and just blatantly deceive for various reasons—were basically drunken encounters in which the purported male victim determined after the fact that it was an encounter he would have preferred not to have and might not have had if he had not been under the influence of alcohol?

The evidence on male and female sexual behavioral predilections that come out of an evolutionary framework makes it hard to believe in significant amounts of female sexual predation of this variety against males in our species without a significant amount of argumentative ammunition to effectively counter the data I have been reviewing here. Whatever one thinks of its status, the distinction between this new category and the typical act of sexual violence against women is stark. Nearly three-quarters of reported rapes of women consisted of victims of forced penetration (ibid.: 18, table 3.1) and not the more amorphous category of alcohol/drug-facilitated penetration. That is, force is nearly always present in male-on-female sexual violence, and it is only rarely claimed to be present in this new category of female-on-male sexual aggression.

On "the Patriarchy"

The evidence of the sex difference is frequently ignored or denied by mainstream feminist-informed sociology—the Munsch article I discussed in the first chapter is indicative of this approach—but sometimes it is acknowledged and given an alternative explanation. "Certainly," this kind of argument runs, "there are some differences in men and women, but all of them are the product of gender socialization and culture that is designed by and for men as a class, in their interests as a class, to keep women as a class in a subordinate position." It is an invisible but powerful structural force—" patriarchy" is the term most often used to name it—that operates in all realms of the life of a society with the sole goal of elevating men and oppressing women. The logic of this claim is the same as that of "systemic or structural racism" in explaining disparities in outcomes and positions for members of racial or ethnic groups. An essential element of social science is leapt over in this kind of approach: The careful determination of cause. The existence of different outcomes for two groups—whether men and women, or members of two racial groups A and B—is by itself only evidence of that different set of outcomes. To demonstrate what caused the different outcomes is a difficult task, but it is the only one that distinguishes a properly scientific social science from journalism, partisan politics, and myriad other non-scientific ways to approach such things.

The idea of patriarchy is defensible scientifically only within a limited purview. In the way it is typically defined by feminist sociologists, that is, as a version of Marxian class-based analysis simply converted from socioeconomic relations to sexual relations, it is belied by voluminous data. Men clearly compete vigorously, even viciously, for sexual access to women, as I have already described above, and much male-caused male suffering in the form of status loss, assaults, and killings results. Some of the male-on-male conflict and violence over females even directly benefits females. In many social systems, especially pre-modern ones, the most significant threat preventing male-on-female sexual violence is serious, even lethal reprisal by male relatives of the assaulted female, especially the father and any male siblings. Careful consideration of empirical details on relations between the sexes in human societies shows little evidence of a class war. It is on the contrary quite clear that what is going on is a complex microlevel game of competition and cooperation involving individual males and females trying to get their reproductive interests realized in a situation wherein overlap of interests between individuals varies and environmental contexts change and present significant obstacles to be overcome.

Yet if we jettison the idea that it is the interests of males as a class that are operative in human society and simply endeavor to understand how the contemporary state of relations of the sexes emerged by individual men and women pursuing the individual reproductive interests they have under natural selection, we can arrive at a compelling explanation of at least some of what feminism tries mostly ineffectually to describe. The logic of sexual selection as I have described it here presents men and women with fundamentally different situations with respect to their reproductive interests. I have already introduced the aspect of the gamete difference and the method by which reproduction happens in mammalian species that has to do with paternity as more uncertain than maternity. Every human mother knows with certainty if a child is hers or not, outside of the possibility that clever thieves may have sneaked into the hospital ward to swap one infant for another. She is present at the birth and typically is immediately in the child's presence. A father, however, can—short of a DNA test—never know with the same degree of certainty if a given child is his offspring. This reality is seldom fully understood in mainstream feminist sociology, and indeed in many cases it seems to be willfully caricatured. Sarah Blaffer Hrdy cites a feminist economist in a telling offhand remark: "We never ask [high-powered] career men whether they have children. I guess we assume some of them don't know" (Hrdy 1999:227). The feminist economist seems genuinely not to have understood that in fact all men do not and cannot know with anything approaching the certainty of women.

In the pure logic of reproductive interest, then, in such a situation, efforts to reduce the uncertainty are quite likely to be made. One of the ways in which, theoretically and in historical fact, men endeavored to reduce this doubt was to exert control over the behavior and whereabouts of female mates when possible. For resource-plentiful men, this might be done by sequestering wives, under protection and observation by others paid or forced to perform this work. For less resource-plentiful men, some greater certainty might be achieved by working in concert with other men to create institutional frameworks—marriage law more restrictive and punitive to women and female sexuality, for instance—that advantage the ability of men to control women and punish those adjudged to have been unfaithful to spouses. Although in the Western world at least political and legal change over the past several hundred year has undone most of those male-advantaged institutional systems, elsewhere in the world they are still rather more in evidence.

The Family and Allomothering

Given the centrality of reproduction to the theory of natural selection, it is no exaggeration to say that the relation of the sexes in our species affects every aspect of human social life, our institutions, and our cultures. One institution, however, is primary in this vision, and this is the family.

There is no more evident place to look to witness how far contemporary sociology has strayed from evolutionary thinking than the direction taken in recent decades by the sociological study of the family. The sole concern in much of this work seems to be to recognize the pluralistic forms in which humans in contemporary individualist societies group themselves into small living groups of mutual concern, which is effectively the new sociological nomenclature for "family." In this blind absorption into the cultural innovations of a particular kind of society at a particular moment in history, the basis of the family in the production and nurturant care of offspring is occluded. Humans are certainly distinguished from many other animals, including our ape relatives, by significant behavioral flexibility with respect to mating and family structure, but this does not mean, as much contemporary sociology infers, that the diversity is infinite or that it is no longer founded on biological universals (Spear 2016). In all human history, and in most of the world still today, the vast majority of families consist of adult parents and the children they have produced, sometimes with older generations of parents or other closely biologically-related individuals also co-habiting with these individuals. Mothers provide significantly more nurturant care for children, though allomothering—the procuring of substitute intensive caregivers for periods of time while mothers attend to other necessary work, most importantly contributing to food production for children—is common. The typical allomother is another female closely related to the mother. Younger potential caregivers are inexperienced and more prone to distraction, while the unrelated are less likely than those whose own inclusive fitness is strongly invested in the involved child to provide care approaching the optimal level. Maternal grandmothers are preferred allomothers across cultures and history.

The basic facts of the emergence of the human family in its baseline form of male and female parents with their offspring can be briefly presented. At the core of the emergence of the human family is the fact that infants in the species *Homo sapiens* are basically helpless at birth, radically incompetent at doing anything in the way of looking after themselves for years afterward. Compared to other primate species close to us, some of which also produce offspring that are comparatively much less precocious than those of other mammals, our babies need more intensive care by others for longer periods of time to make it to self-sufficiency. In mammals, females carry fetuses to term and give birth to them, and they also lactate sufficiently to nourish offspring early in life. The mother–child relationship is therefore typically the strongest bond across a wide range of mammalian species, including our own. Lactation is the single factor most powerfully driving greater motherly commitment to nurturant care (Hrdy 1999:122). It is an entirely female phenomenon in the animal kingdom, apart from one species of fruit bat. The evidence indicates that it may have evolved from something like a plugged sweat gland that exuded nutrients that an infant cuddling near the mother might have ingested and then gained some nutritional and immunological benefits. Over evolutionary time, the sweat glands became teats. The neurochemical benefits of oxytocin are generated for mothers by breast-feeding. Culture on top of this can of course further solidify the pattern, but it is produced from deeper facts driven by biology and natural selection.

Biological kin have ways to innately recognize shared/inclusive fitness, and they will readily sacrifice to help kin when they will not do so to help non-kin. All early human societies were almost certainly relatively closely related kin groups. The biological parent–offspring relationship is characterized by greater shared fitness (approximately 50%) than any other relationship except identical twins (approximately 100%) and full siblings (approximately 50%), and so biological parents can typically be counted on to be more concerned for the welfare of their own offspring than others who are not as closely biologically related to them. Concern and investment in given infants can be expected, all other things equal, to decrease as genetic distance between the infant and the potential caregiver increases. From these biological facts, significant advantages arise for such creatures to bond together into familial units based on biological relatedness, and for male and female parents of offspring to be central players in that familial unit.

So much for the basics. Now for some complexities. Sarah Blaffer Hdry performs the invaluable task of attending fully to the ways in which sexual selection informs us as to how female humans make choices in the game of reproductive competition. She describes "a new view of mothers," unvarnished by the mythology about pure, self-sacrificing maternal goddesses and attentive instead to real women making decisions in the real world. The basic computational issue, which the gamete difference alters but does not eliminate, for females is the same as for males: Do you elect to produce many offspring with little investment in each, or do you produce only a few, with significant investment in each one (Hrdy 1999:32)? The working out of this calculus depends on myriad environmental factors, and Hrdy musters a great number of examples in humans and in other species to show how it works. Some mothers will abandon offspring when it becomes apparent that further investment in them is virtually guaranteed to be wasted effort. Female mice can reabsorb a fetus when they smell an unfamiliar male in the area because they are genetically programmed to understand that such males are a lethal threat to offspring they have not sired (ibid.:89).

Perhaps the key element of human mothering and family that Hrdy discusses is that of allomothering. From the first in the life of our species, mothers needed to forage to feed themselves and offspring, and such labor takes time and energy that cannot always be shared with childcare, so reliable allomothers are actively sought. They are not always available, and in such situations, mothers make hard decisions—perhaps foregoing foraging, perhaps abandoning offspring, perhaps trying to compromise the two activities by doing both at suboptimal levels. Ultimately, only the real mother is likely to defend offspring with maximal ferocity—this is genetic in large part, driven by hormonal enhancement of bonding and aggression—but sacrifices can be made in some situations, and imperfect allomothers are sometimes better than no allomother at all. The calculus is delicate and contingent on many factors (ibid.:152).

However important the gamete difference is to understanding reproductive strategies in males and females, it is easy to overestimate its role if we do not account for important contextualizing factors. If for example you compare male and female sex drives, the former is consistently higher if we look at the entire calendar year. But this overlooks a difference in fertility for males and females. The former can produce offspring any time an ejaculate can be produced. Females, however, can only reproduce during the few days around mid-phase of their menstrual cycle. If you compare male sex drive to female sex drive only during their fertile periods, the two are much more similar. The heightened mid-cycle sex drive in women is also responsible for seemingly confounding phenomena such as the greatly increased attraction women have in this time frame for men who give off the most obvious cues of masculinity but who are likely to be relatively lacking in compassion and commitment potential, i.e., men who are not likely to be optimal partners in the altered environments in which modern humans live but who would have been more selected for in the Pleistocene environment in which our psychologies were largely built (ibid.:225–226).

The preceding body of information is typically overlooked in introductory texts in the sociology of the family, and many of the assertions accepted as established in that field today are revealed to be suspect in light of such facts. It is sometimes claimed, for example, that the rise in female-income headed families is an indication that the gendered division of labor that existed through most of human history—fathers laboring outside the home, while mothers attended primarily to labor inside the household—has given way to a new pluralism in the ways in which mothers and fathers divide familial labor. The evidence, it is argued, has to do with the fact that some 40% of families in the contemporary US have a female as the main breadwinner. This is true, but the reasons for it are not so much that the older gendered division of labor has fundamentally shifted as that the single-parent family—overwhelmingly female-headed—has become such a common feature of modernity. A full two-thirds of that 40% of female-income headed families are single parent families, and these families skew heavily toward the lower income brackets. In the vast majority of male–female two-parent families, the venerable sexual division of labor is still much in evidence.

It is also sometimes suggested in this literature that a significant number of women are rejecting marriage and children altogether as inconsistent with their life and career plans. Survey data, however, indicate that around 95% of American women express a desire to marry and to have children at some point, a figure that has not changed significantly in the past quarter century or more. A more careful look at the dating and mating markets facing young Americans shows not that women or men have willingly turned against families with marital partners and their biological children. Rather, the growing portion of the first third of life dedicated to education and career advancement, along with the liberalization of accessibility of sexual activity outside marriage, have led to a postponement of marriage until later in life. This development has—because of the facts of the sex difference—inevitably meant that some number of the women who wait until closer to the closing of the biological window for reproductive possibility faced by females of our species end up unable to procure mates at all unless they are willing to choose males who are sub-optimal in their ranking schemas. Increasing numbers of males are also opting to leave the marriage market altogether, given that the possibilities for sexual activity are now so much more plentiful outside marriage than they were in the past (Regnerus 2017). Attitudes about a more or less traditional family model show little evidence of having substantially shifted. Other changes in American society have simply produced an alteration in the field of possibilities for individuals that make it more difficult for some of them to achieve their desires and that allow others to satisfy the behavioral mechanism of reproduction outside of the familial institutional form that formerly encompassed it entirely.

Much remains unknown concerning the ways in which fundamental changes to the nature of families will affect children. When the genetic/biological difference between the alloparent and the infant grows, however, we reasonably expect by the logic of Darwinian theory for the risk of less optimal care to increase to some degree. A significant amount of data suggest that such risks do increase for children in families where stepparents, and especially stepfathers, are present. Daly and Wilson (1999) refer to this as the "Cinderella effect." A stepparent is more likely than a biological parent to have entered the mating relationship without a great deal of enthusiasm about the role of parenting of the child/ren involved. Such mates are more likely to see their relationship with the child as one of competition for the time and resources of the biological parent of the child, though this dynamic exists in biologically intact families as well—again, conflicts of interest are endemic in human society, and the family is not an exception to this rule. Two adoptive parents genetically unrelated to the child are typically individuals with a strong interest in parenting but who have been denied the opportunity to be biological parents for various reasons, so we might perhaps expect something of a correction of the "Cinderella effect" risk to children in such families. In general, though, the expectation in the theory of natural selection is that we should have as our default expectation that there will be an observable difference in outcomes for children as a group in families in which adult members not biologically related to the children are present.

In modern societies, alloparenting now includes all sorts of still more distant relations—e.g., daycare workers who act as parents for much of the day for many children from an early age. In some settings, e.g., upscale middle class couples in which both adults are fully engaged in time-consuming careers outside the family, daycare workers, babysitters, and other paid alloparents may actively parent and spend time with children as much as or more than biological parents. Here too there is significantly higher risk of suboptimal care than in parenting by biological parents. It is important to note that saying risk increases with stepparents and other alloparents does not tell us anything about the magnitude of the additional risk, and in fact we find in societies like ours that serious abuse of children is comparatively rare. Even when we count abuse such as neglect, it is a matter of perhaps 1% of all children at most who experience this (US Department of Health and Human Services 2016). So even though stepparent rates of abuse are considerably higher than rates of biological parents, it is still true that this involves a small minority of stepparent families because the baseline rate is so low.

Another Object Lesson in the Weakness of Modern Biophobic Sociology of the Family: Homosexual Parenting

Despite the uncertainties about how things will go in the new world of pluralistic families, mainstream sociology speaks confidently about how much research we have to show that nothing will change—beyond the positive goal of greater diversity—as we move increasingly into a cultural space that does not privilege or particularly encourage family units based on biological relationship of both parents and the involved children. But when one looks closely at the evidence for the claims made, questions emerge quickly. The issue of marriage involving two homosexual parents and their children—whether adopted by both or related biologically to one of the parents, from previous relationships or other methods—is of considerable interest on this topic to social scientists and policy makers. In 2001, the sociologist of marriage and family Steven Nock produced an exhaustive affidavit in the hearing by the Canadian Supreme Court of the case that ruled against the then-existing common law definition of marriage. In it, Nock demonstrated how little can be scientifically inferred from the dozens of studies—all with small samples and unrepresentative of the national population—on the topic that were then typically cited by advocates as evidence that the debate had been settled and that there were no significant differences in outcomes for children of biological and homosexual parents. In good scientific form, Nock did not assume what could not be known and limited himself to advising the court that the case on whether or not homosexual families yielded the same outcomes as heterosexual ones do was at that point not settled, contrary to the claims of advocates that it was. We would, he argued, have to wait for more rigorous research to know the answer. Nock's affidavit was hostilely attacked by another affidavit written by two other sociologists, Judith Stacey and Timothy Biblarz, which unwittingly indicted more or less the entirety of the kind of sociology they practice by admitting that Nock's criticism of the existing studies was accurate but then affirming that this was the state of the art in the entire discipline.

Consistently since Nock's affidavit, other advocates of gay marriage have repeated the claim of no differences in outcomes, sometimes without any new evidence, but sometimes arguing that newer studies have answered to Nock's objections. Exemplary here is a kind of American equivalent of the Nock affidavit in reverse, a 2014 American Sociological Association summary of research submitted as an amicus brief supporting proponents of the federal recognition of gay marriage as a civil right (Manning et al. 2014; referred to hereafter as "Manning"). A decade and a half after Nock, the state of the research had changed somewhat, but descending below the article's rhetorical sleight of hand and engaging in careful consideration of the studies cited as indisputable evidence of the claim makes it clear that the situation is still scientifically unsettled.

The Manning study claimed to show more than 40 studies that support the "no differences" hypothesis specifically in an American context, but a glance through their bibliography shows that many of the studies they cited do not fit this definition. They included, for example, five different studies by the same group of authors, all of which are studies of same sex parenting in the Netherlands. One of these (Bos and Gartrell 2010) has to do with stigmatization of children in lesbian families but does no comparison to children in heterosexual families. Another (Bos et al. 2008) compares lesbian families in the US to lesbian families in the Netherlands; it does not compare either to heterosexual biological families in either country. Still another (Bos and Gartrell 2011) compares children in lesbian families conceived with a known donor to those in lesbian families conceived with an unknown donor; again, no comparison at all to heterosexual families. There is simply no scientific way that these studies can be understood to say anything about the comparative analysis of child outcomes in biological and homosexual families. Numerous other studies they cited have similar problems.

Many of the studies cited fall victim to the same methodological criticisms raised by Nock in 2001. Studies done using convenience samples are of little use for answering the question of whether there are differences in outcomes for children in same sex and heterosexual families. Large and nationally representative samples are needed to do that. As Nock pointed out, very few studies of same sex families are based on the kinds of statistically representative and large samples that we need to make firm scientific judgments about differences with respect to heterosexual families. In 2001, when Nock wrote, there were basically no such studies. In 2014, the time of writing for Manning, there were only a handful. They cited five: Allen et al. 2013, Regnerus 2012a and 2012b, and Rosenfeld 2010 and 2013. As they admitted, many more of the studies they cited use the National Longitudinal Lesbian Family Study, which consists of interviews with "donor-inseminated lesbian mothers five times from insemination or pregnancy to the child's 17th birthday." This sample is without doubt unrepresentative of same sex families generally because it consists only of lesbian families and does not consider homosexual male families. It also certainly skews toward wealthier lesbian couples, as artificial insemination is expensive and only a minority of states have requirements for insurers to offer any coverage for it, i.e., many insurers do not cover it.

Once one has narrowed Manning to the handful of studies that are at least based on samples that offer some hope of scientific conclusions, what remains? Two studies (Rosenfeld 2010 and 2013) that broadly support Manning, and three (Allen 2013 and Regnerus 2012a and 2012b) that do not. But in fact, only one (Rosenfeld 2010) supports Manning, because the other is not a new study—it is only Rosenfeld's response to a methodological criticism of his earlier study. The author of that critical response to Rosenfeld was the study by Allen. If we count that as a new study—it creates a new statistical model for the original data, which enables a significant difference to emerge showing that children in same sex families do substantially worse in school than children of heterosexual biological families—this leaves us with one study supporting Manning and three against it. If we throw the Allen article out too, then we have one for the thesis of Manning, two against. If we want to be picky and note that one of the Regnerus articles (2012b) is actually rather akin to the Allen study insofar as it is a relatively slight tweaking of the original study on which it is commenting, then we have one study for Manning, one against. That looks rather less like a clear scientific vindication of the argument that there are no differences in outcomes for children in biological and homosexual families than is claimed by the authors.

The Manning study offers much criticism of the Regnerus study on homosexual parenting, which was widely attacked in the mass media as well because it found just what an evolutionary theory would suspect to be the case, that is, that children in families in which adults who are not their biological parents are present suffer deficits compared to children in biological parent families. They fundamentally misrepresent the Regnerus study's soundness and its claims. Regnerus has rebutted most of the criticisms in one of the very articles by him that they cite (Regnerus 2012b). He admits that his study has limitations, as all scientific research does. What is important here is not so much whether one thinks there are problems with the Regnerus study. It is the remarkable fact that the Manning study never once mentions that the one statistically representative study with a large sample in their list that supports their claim about the entirety of the literature (Rosenfeld 2010) was criticized in similarly strong ways by another researcher (Allen 2013), who showed that Rosenfeld's conclusions were debatable. So how effectively has Manning shown readers the scientific lay of the land on this question, if they have not even been honest and even-handed enough to note this?

On this complicated question, then, as on some others described in this chapter, we can reasonably say the jury is still out. Enough is known, however, about how the sex difference matters in our species, at virtually every level of behavior and social organization, to place one's bets with some degree of confidence—while still, of course, being fully committed to doing the science and letting the truth out, whatever its form—regarding what the verdict will look like when it comes in.

References

Allen, D. W., C. Pakaluk, and J. Price. "Nontraditional Families and Childhood Progress Through School: A Comment on Rosenfeld." *Demography*, 50 (2013):1–7.

Archer, John, ed. *Male Violence*. London: Routledge, 1994.

Baron-Cohen, Simon. *The Essential Difference: Forget Mars and Venus and Discover the Truth about the Opposite Sex*. New York: Penguin, 2003.

Bos, H.W., and N. Gartrell. "Adolescents of the USA National Longitudinal Lesbian Family Study: Can Family Characteristics Counteract the Negative Effects of Stigmatization?" *Family Process*, 49, 4 (2010):559–572.

—. "Adolescents of the US National Longitudinal Lesbian Family Study: The Impact of Having a Known or an Unknown Donor on the Stability of Psychological Adjustment." *Human Reproduction*, 26, 3 (2011):630–637.

Bos, H.W., N. Gartrell, F. van Balen, H.H. Peyser, and T.M. Sandfort. "Children in Planned Lesbian Families: A Cross-cultural Comparison between the United States and the Netherlands." *American Journal of Orthopsychiatry*, 78, 2 (2008):211–219.

Brownmiller, Susan. *Against Our Will: Men, Women and Rape*. New York: Simon and Schuster, 1975.

Churchland, Patricia. *Touching a Nerve: Our Brains, Our Selves*. New York: W.W. Norton & Company, 2013.

Daly, Martin and Margo Wilson. *Homicide: Foundations of Human Behavior*. New York: Routledge, 1988.
—. *The Truth about Cinderella: A Darwinian View of Parental Love*. New Haven: Yale University Press, 1999.
Del Giudice, Marco, Richard Lippa, David Puts, Drew Bailey, Michael J. Bailey and David Schmitt. "Mosaic Brains? A Methodological Critique of Joel et al." *Social Science Research Network*, (Dec. 23, 2015).
Epstein, Cynthia Fuchs. "Similarity and Difference: The Sociology of Gender Distinctions." Pp. 45–64 in Janet Saltzman Chafetz, ed., *Handbook of the Sociology of Gender*. New York: Springer, 2006.
Fausto-Sterling, Anne. "The Five Sexes: Why Male and Female are not Enough." *The Sciences*, (March/April 1993): 20–24.
—. "The Five Sexes, Revisited." *The Sciences*, (Jul./Aug.2000).
Fox, Richard, Jennifer Lawless and Courtney Feeley. "Gender and the Decision to Run for Office." *Legislative Studies Quarterly*, 26, 3 (Aug., 2001): 411–435.
Freeman, Scott and Jon Heron. *Evolutionary Analysis*, 4th edn. New York: Pearson, 2007.
Frodi, Anne, Jacqueline Macaulay and Pauline Thome. "Are Women Always Less Aggressive Than Men? A Review of the Experimental Literature." *Psychological Bulletin*, 84, 4 (1977):634–660.
Goldberg, Steven. *The Inevitability of Patriarchy: Why the Biological Difference between Men and Women Always Produces Male Domination*. New York: William Morrow and Company, 1973.
Hrdy, Sarah Blaffer. *Mother Nature: A History of Mothers, Infants, and Natural Selection*. New York: Pantheon Books, 1999.
Joel, Daphna, Zohar Berman, Ido Tavor et al. "Sex Beyond the Genitalia: The Human Brain Mosaic." *PNAS*, 112, 50 (Dec. 15, 2015):15468–15473.
Kelly, Robert. *The Lifeways of Hunter-Gatherers: The Foraging Spectrum*. New York: Cambridge University Press, 2013.
Kortüm, Hans-Henning and Jürgen Heinze, eds. *Aggression in Humans and Other Primates: Biology, Psychology, Sociology*. Berlin: De Gruyter, 2013.
Lassek, William and Steven Gaulin. "Costs and Benefits of Fat-free Muscle Mass in men: Relationship to Mating Success, Dietary Requirements, and Native Immunity." *Evolution and Human Behavior*, 30 (2009):322–328.
Lawless, Jennifer and Richard Fox. *It Takes a Candidate: Why Women Don't Run for Office*. Cambridge: Cambridge University Press, 2005.
Manning, Wendy, Marshall Fetro, and Esther Lamidi. "Child Well-being in Same-sex Parent Families: Review of Research Prepared for American Sociological Association Amicus Brief." *Population Research Policy Review*, 33, 4 (Aug. 1, 2014):485–502.
Nock, Steven L. Affidavit for *Halpern et al., v. The Attorney General of Canada.*
Regnerus, Mark. "How Different are the Adult Children of Parents who have Same-sex Relationships? Findings from the New Family Structures Study." *Social Science Research*, 41, 4 (2012a):752–770.
—. "Parental Same-sex Relationships, Family Instability, and Subsequent Life Outcomes for Adult Children: Answering Critics of the New Family Structures Study with Additional Analyses." *Social Science Research*, 41, 6 (2012b):1367–1377.
—. *Cheap Sex: The Transformation of Men, Marriage, and Monogamy*. New York: Oxford University Press, 2017.
Ritchie, Stuart J., Simon R. Cox, Xueyi Shen et al. "Sex Differences in the Adult Human Brain: Evidence from 5,216 UK Biobank Participants." *Cerebral Cortex*, 28, 8 (Aug. 2018):2959–2975.

Rosenfeld, M.J. "Nontraditional Families and Childhood Progress Through School." *Demography*, 47, 3 (2010):755–775.
Rosenfeld, M.J. "Reply to Allen et al." *Demography*, 50, 3 (2013):963–969.
Roughgarden, Joan. *Evolution's Rainbow: Diversity, Gender, and Sexuality in Nature and People*. Berkeley: University of California Press, 2009.
Sax, Leonard. "How Common is Intersex? A Response to Anne Fausto-Sterling." *The Journal of Sex Research*, 39, 3, (Aug. 2002):174–178.
Sigal, Pete. "Imagining Cihuacoatl: Masculine Rituals, Nahua of Goddesses, and the Texts of the Tlacuilos." Pp. 12–37 in Kevin Murphy and Jennifer Spear eds, *Historicising Gender and Sexuality*. London: Wiley and Sons, 2011.
Smith, J. Maynard. *The Evolution of Sex*. New York: Cambridge University Press, 1978.
Smith, Sharon G., Jieru Chen, Kathleen C. Basile, Leah K. Gilbert, Melissa T. Merrick, Nimesh Patel, Margie Walling, and Anurag Jain. *The National Intimate Partner and Sexual Violence Survey (NISVS): 2010–2012 State Report*. (April 2017).
Spear, Rebecca. "Beyond the Nuclear Family: An Evolutionary Perspective on Parenting." *Current Opinion in Psychology*, 7, (2016):98–103
Stacey, Judith and Timothy Biblarz. Affidavit in response to Nock affidavit.
Stemple, Lara and Ilan Meyer. "The Sexual Victimization of Men in America: New Data Challenge Old Assumptions." *American Journal of Public Health*, 104, 6 (June 2014):19–26.
Stoet, G., and D.C. Geary. "The Gender-equality Paradox in Science, Technology, Engineering, and Mathematics Education." *Psychological Science*, 29 (2018):581–593.
Symons, Donald. *The Evolution of Human Sexuality*. New York: Oxford University Press, 1979.
Trivers, Robe rt. *The Folly of Fools: The Logic of Deceit and Self-Deception in Human Life*. New York: Basic Books, 2011.
United States Department of Health and Human Services. *Child Maltreatment 2016*.

6 Stratification, Status, and Inequality in *Homo sapiens*

A *New Yorker* story from January 13, 2020, titled "The Equality Conundrum," makes a statement about inequality in its byline that could stand as the summary statement on the topic by the sociological mainstream: "We all agree that inequality is bad." A few paragraphs in, we are told that "a plurality" (27%) of Americans ranked it the single greatest danger facing the world. We learn also that of the world societies that rank inequality as the most important of a number of problems—others included religious and ethnic hatred, AIDS and disease, pollution, and nuclear weapons—all but one, Argentina, are in Western Europe. In poorer parts of the world, other problems outweigh inequality. The piece gets in a gratuitous kick or two at President Trump, perhaps the most telling of which gets at another deep aspect of widespread understanding about inequality in the West. He is quoted, years before becoming president, saying that the idea that we are born equal, as stated in our Declaration of Independence, is subject to empirical criticism, whatever its importance as a statement of democratic political philosophical first principles: "Some people are born very smart, some people are born not so smart. Some people are born very beautiful, some are not."

Stratification as Unethical Greed: The Mainstream View

I often give students an article by the psychologist Paul Piff to illustrate some of the problems with the mainstream social scientific understanding of stratification, status, and inequality. Piff has written a good deal on the topic over the years and his work has gotten significant and congratulatory coverage in the mainstream press. In the short piece I give students, "Higher Social Class Predicts Increased Unethical Behavior," the argument is right there in the title. Piff finds in a summary of seven studies that those higher up in the socioeconomic order are more prone to behaving unethically, which in his terms means "self-interestedly." The article is predicated on a mountain of assumed, unargued moral philosophy that is fundamentally at odds with the basic evolutionary theoretical understanding of human interests and moral systems outlined in chapter four. It is claimed that "greed…is a robust determinant of unethical behavior," and Plato and Aristotle are cited as sources to sustain the idea that greediness is "at the root of personal immorality…driv[ing] desires for material gain at the expense of ethical standards" (Piff, Stancato, Côte et al. 2012:4086, referred to hereafter as "Piff").

Methodologically, the studies are profoundly suspect. Several are field studies in which observers took notes about cars at four-way stop intersections and how they behaved in their following of pecking order rules for turn-taking. The social class status of subjects was determined by the make of their automobile. The article does not give the status code for make and model of automobiles used in the studies, but this is a leaky method for determining objective class standing at best. It is well known that conspicuous consumption is a relatively frequent phenomenon in contemporary American society to give a status impression above the actual position occupied in the class hierarchy. Many wealthy individuals engage in an inverse process, avoiding purchase of luxury consumer items to escape unwanted critical and potentially criminal attention. The Piff article makes no reference at all to these phenomena. In other studies discussed in the article, social class for subjects is determined solely by their self-definition. It is well-established that people typically do not have good objective knowledge of where they fit in the class hierarchy, and there are numerous forces—especially the desire not to be seen as too poor or too rich—that lead people to purposefully mislead about this aspect of their identity. In another study Piff describes, the "unethical behavior" was operationalized as the taking of candy by subjects from a jar that they had been told was for children. In the index summary of the study, but not in the main body of the text, it is revealed that subjects had in fact been told that they could help themselves to the candy. How unethical is it then to take candy that you have been told you may take? In yet another study, the unethical behavior consisted of cheating in a dice rolling game to win a prize—the prize was a retail gift certificate of a small monetary amount. This study, like many others of a laboratory nature that operationalize some broad behavioral tendency in an artificial manner that is evidently distant from real life scenarios, is likely not telling us a great deal about how greedy or altruistic subjects are.

But the methodological problems of the Piff study are not my focus with students. Instead, I ask them what defensible theory of human nature enables us to imagine not just that poorer people are more altruistic than wealthier people (an explicit claim made in the article), but more broadly to believe that human behavior that is "self-focused" is somehow aberrant and out of sync with the normal default behavioral predilections of *Homo sapiens* (ibid.:4086). They will frequently talk about how it just seems obvious to them that those higher up on stratification ladders behave more self-interestedly—indeed, they will sometimes add, that is the only possible explanation for why they are there in the first place—but they seldom have thought about the matter sufficiently to be able to robustly defend that claim.

In fact, as researchers critical of Piff's findings have indicated, social class position is a far more complicated variable than indicated in this study. Numerous measures can be used to define it—financial wealth, income, job type, permanence of employment, supervisory position at work, high education level—and these variables do not consistently line up to produce a uniform, discrete upper class. More, different measures of status correlate with different attitudes and behaviors. High status in wealth, job type and education are typically correlated with higher levels of volunteering, an altruistic quality according to Piff. Political attitudes, and presumably voting and other ethical behaviors associated with same, also vary in the higher status groups. High status in income, job type and education predispose subjects to align with the political left, which also rubs against Piff's analysis. Better designed natural and experimental studies indicate that there is little difference in levels of unethical, i.e., self-interested behavior by higher and lower status groups. Instead, we find that all groups are prone to act in a self-interested manner, given the appropriate cost/opportunity situations. If the odds of being caught are low, and the potential payoff is significant, most people of all status groups will do so (Trautmann, Kuilen, and Zeckhauser 2013).

At root, why is there status inequality and stratification in human societies? What basic elements of human nature produce and explain the existence of this aspect of human social organization? The conventional argument in sociology is the same one dominant in much of the media and popular culture, the same one presumed *a priori* in the Piff article: Some small subset of humans illegitimately and unethically pursue their own self-interest inordinately and without consideration of others, they rise to positions of dominance over others, and they then mercilessly defend the unjust position they have attained through their own greed against righteous subordinates who have been unfairly denied their moral desserts.

A scientific understanding of how status works and why human societies are so demonstrably tied up with the phenomenon would do well to bracket the evident moral self-righteousness of this mainstream position. Whether inequality is to be criticized and condemned, and even rejected as altogether unacceptable, all of this has nothing whatever to do with the project of accurately understanding it. It is a telling fact that mainstream sociology has produced so very little in the way of an effort at testable theories and has instead contented itself with moral condemnation and gesturing at the utopian condition, never realized anywhere in practice to this point, of a social order without stratification.

Nearly every introductory text on sociology presents hierarchy and stratification as problems to be solved by proper knowledge and an egalitarian political system. A representative such text (Schwalbe 2007) describes how the author starts the discussion of inequality in his classes with what he calls "the paper plates game." This involves ten students and ten paper plates that are given out to the students proportionally to US income inequality, so that the plates are distributed quite unevenly. Students are shocked, the author writes, at the level of the inequality. At the base of the lesson is the assumption that each student deserves and should get exactly one plate. Nowhere does the author take on the difficult philosophical question of why that should be the expected or even the most desirable result, given what we know about how human societies and human nature operate. A properly scientific account of this topic would start, by contrast, with the recognition that inequality and stratification are features of human societies we find everywhere, without exception. Even in developmentally early societies in which economic egalitarianism rules, individual inequality by status and stratification along lines of sex and age are essentially universal. Why should this be so? Are all societies distortions of what they "ought" to be? The starting point for a properly social scientific position on these topics is simply to recall that the moral and the scientific elements of this and other similar topics must be separated. What one thinks in terms of good and bad about status systems in human societies has nothing to do with what the evidence shows us about their place and frequency in human social order or which theories of human nature most adequately explain that evidence. Indeed, there might well be something interesting to learn about the true nature of hierarchy in human societies in the fact that so many individuals in modern societies react to its existence with this kind of emotionally-charged indignation and outrage.

The founders of the discipline demonstrated more willingness than we find in contemporary sociology to take stratification on scientifically. Weber was skeptical of the idea, already spreading in the intellectual classes in the Germany of his day, that radical social levelling was the appropriate way forward. A central aspect of his own theory had to do with a rejection of the Marxian notion that an objective position in the social class hierarchy necessarily leads to a class consciousness that, in the case of Marx's proletariat, would inevitably point to a critique of the existing hierarchy and a social revolution. Weber is the figure in the history of sociology credited with most forcefully pointing out the limits of the Marxian view of stratification as a dynamic of class groups against other class groups. Whatever the objective similarities of the socioeconomic positions of two members of a purported social class, there is no inevitability to the emergence of a shared sense of the existence of that class and their membership in it. And history to this point has given us no reason whatsoever to believe in the necessary, or even likely, movement of human history toward a solution to the problem of stratification derived from a revolutionary victory by a social class that will in that victory choose to do what no other ruling class has ever done, that is, voluntarily give up their rule and usher in a social order without hierarchy. Durkheim took it for granted that even a functional and healthy division of labor would produce inequalities, and he was not troubled that such a development posed an insuperable problem that had to be eliminated. This was in his view an inevitable outcome of the fact that not everyone is equally suited to do every task, and accommodating individuals to their objective place in the division of labor is an important task for moral society. Though Marx was obviously led astray from scientific analysis of stratification in the politically prophetic aspect of his writing, he too contributed useful elements to a naturalistic, objective theory for understanding how hierarchy emerges in human history as an inevitable consequence of our effort to dominate the natural world through labor, which necessarily extends—and especially in certain periods of human history—to domination of one another.

The real scientific attention to hierarchy and stratification in the founder's generation, however, is to be found in the small, largely forgotten group of Sumner, Spencer, and Pareto I discussed briefly in the first chapter. Each of these attempted a naturalistic approach to the phenomenon, careful to avoid crude moralizing in place of analytic rigor, and each started from a theory of human nature that was more independent of the Kantian and Hegelian idealism and moralism that was the fundamental philosophical force driving Durkheim, Weber, and Marx. But, as I indicated in my earlier discussion of them, the promise of their work did not survive the emergence and increasing dominance of the critical sociological perspectives of the mid-twentieth century.

Stratification in Historical Relief

What are the bases of a scientific sociological theory of stratification? One of the first things that must be said as we take on this topic is that a comparative angle is the only one that makes the slightest bit of sense in scientific terms. If we begin with moral judgments and prescriptive opinions about this phenomenon as it appears in modern industrial societies, absent a view of how it compares and is related to stratification in earlier societal types to guide our understanding, we can be certain that we will fail to inquire sufficiently deeply into the phenomenon. The lack of an historical perspective means that, in my own experience at least, students almost never have the comparative perspective required to fully understand levels of stratification in contemporary societies.

Viewed in the proper historical relief, stratification takes on a rather different hue than it acquires in nearly all contemporary sociology. Everything we know about hunter–gatherer societies, the societal stage in which humans have spent perhaps 95% of the time we have been on the planet as a species, suggests that hierarchy and inequality existed but that it was rather less in evidence or even functionally non-existent on the axis in which it is dominant in most contemporary societies.[1] There is considerable variation in this societal stage of human development (see e.g., Kelly 2013 for fascinating detail), as indeed is true in later stages too, but material wealth disparities were almost certainly relatively low in simple hunter–gatherer societies. A good piece of the logic of this case rests on the fact of the relative lack of economic prosperity of such societies. These are the kinds of human societies that were closest to what we can properly call subsistence societies, in which there is essentially no economic surplus and the group must acquire all its food daily. Nomadism is also a typical feature of such societies, which means that any personal goods that cannot be frequently picked up and hauled off to a new site are excessive. Many such societies establish informal norms for effectively policing those who insist on overly egotistic accumulation or hoarding, especially of food. Hunters who do not share the catch are often denied a share from hunts in which they do not make kills, and they may face more stringent disciplinary mechanisms, leading all the way up to expulsion or even death, if they do not appropriately adjust their behavior. The danger in a small subsistence society of hoarding and excessive material inequality would have been transparently obvious to everyone. At the most basic level, the resentment of the deprived would have been rather more difficult for elites to shield themselves from than it is in modern mass societies. At the maximal level, there would potentially have been a real threat to the material existence of poorer members of the group and by extension of the entire group (Flannery and Marcus 2012). This is however not to say that other forms of inequality did not exist in hunter–gatherer groups, or that such societies were, as is frequently claimed in the popular press and indeed by some social scientists, free of hierarchy. Such stratification is based not on wealth but on prestige, and it is often narrowly based on individual characteristics and at least to some degree meritocratic. Few or no barriers of birth prevent an individual with talent for hunting or resolving conflicts in the group from reaching a position of greater prestige than obtained by anyone else in his immediate family.

Material inequality, up to and including the existence of slavery, is already in evidence in more complex hunter–gatherer societies such as those of some of the indigenous peoples of the American Pacific Northwest Coast (Kelly 2013:241–268). But it is the invention of horticulture, which makes possible a significant surplus, that presents in human history the moment of the emergence of significant levels of inequality in wealth. As soon as a surplus is made possible by the new and more efficient mode of food production, interest in effectively controlling it arises among existing elites. Permanent settlements arise here, and this makes it possible for wealthy individuals to accumulate stockpiles of possessions they could not possibly have carried with them in the nomadic period of human life. Kinship systems remain in horticultural societies the social organizational principle of prime importance. These systems commonly extend broadly and connect significant numbers of people. Effective clan leaders emerge as *de facto* heads or chiefs of substantial groups of extended kin, and they are able through their political power to accumulate wealth. Warfare, relatively rarer and more restricted though not unknown in the hunter–gatherer period, begins to become more common as horticultural societies grow and come into increasing contact with one another. The emergence of material wealth surpluses in stored crops and especially in domesticated animals makes possible a rise in collective theft as a potentially easy way to grow surplus at the expense of other groups. There is evidence here of a growing cult of the warrior, with men formerly occupied in the hunt and now freed from that time-consuming activity by increased horticultural production using their time to hone some of the same physical capacities useful in hunting for application in fighting with other men in military engagements. Successful warriors and leaders of warriors can also enrich themselves.

As horticultural societies grew and horticultural technology developed, eventually leading to the advent of the harnessing of animal energy and the use of the plow in farming, which massively increases productivity, inequality rises rapidly. Slaves, often captured in military expeditions, become a common feature of the human landscape. Increasing stratification in agrarian societies is visible particularly along three axes: That between the small governing class and the large mass of those with no real political decision-making power, that between the small number living in emerging urban areas and the masses of those living in the country, and that between the tiny literate minority and the overwhelming majority of those who were illiterate. It is in the later agrarian period that we see the full-blown emergence of what we today know in sociology as class groups, that is, discrete social categories of people defined fundamentally by their economic activity and the concomitant amount of material wealth this allows them to command. The great majority of the individuals in any agrarian society were peasants and expendables, those directly engaged in agricultural labor in the country and those who performed the lowest menial chores in more urban areas. They typically controlled little wealth and had no direct political power at all beyond the capacity to express outrage through collective violent uprisings and riots, which were mercilessly put down by military forces commanded by elites as soon as they threatened social order. We should avoid using the moralizing language of the post-French Revolutionary democratic partisans and social philosophers to exaggerate the suffering of peasants. They were certainly far poorer than the aristocrats who owned the lands they worked, and inequality that separated the two groups was social and cultural as well as economic, but the evidence of recognition of an organic relationship between these two social strata is great, as Alexis de Tocqueville, Pierre Gaxotte and other historians of the Ancien Régime have documented. The governing class was vanishingly small and separated by a wealth gap from those lowest on the economic hierarchy by a distance even greater than in contemporary industrial societies. Other classes that occupied positions between the two extremes were artisans, priests and other officials of the churches, retainers, and a growing and energetic merchant class, some of the members of which might have been wealthier than some in the governing class.

The emergence of the industrial societal system altered stratification again. The number of different axes increases still more here, and levels of inequality remain considerably higher than they typically were in most horticultural and all hunter–gatherer societies. Yet several changes in culture and in the efficiency of technology translate into a relative decline of levels of inequality in comparison to the late agrarian period. Democratic discourses of human rights—generally based in universalizing religions that begin to emerge in this epoch of human history—and much greater participation and representation of all members of industrial societies lead to moral outrage at some of the institutions and practices that produced massive inequality and suffering in agrarian societies. Slavery is outlawed, a great deal of the most egregious economic exploitation is undone by legal and political means, and the expendable class is eliminated by social welfare programs and human rights law. Technological improvements in food production and medical care translate into an objectively superior standard of living for everyone in industrial societies, including those at the very bottom.

Inequality as Result of Natural Selection

The theory of human interests given in chapter four illuminates human society as an innately competitive endeavor in which the involved competitors never perfectly share interests. Human diversity—along the sex divide, most fundamentally, but operating in myriad other ways as well based on the natural distribution of individual characteristics in any population—reveals that the competition will typically not be between precise equals, and therefore superordinates and subordinates are an almost certain outcome of most competitions. The evidence supporting the claim of the basic natural fact of unequal attributes and the consequences that frequently has for status is so overwhelming that it takes someone laboring under the illusions of a powerful belief system indeed to overlook it.

Let's return to the statement from Donald Trump cited at the outset of this chapter as an illustration. The question of the meaning of Jefferson's statement in the Declaration in political theoretical terms is one we can safely put aside in considering Trump's statement in light of what we know from the scientific study of the distribution of attributes in the human population. That some people are generally viewed as more attractive than other people is a fact every member of every society knows intuitively as a competent member of his society. Cultural relativists frequently argue that different cultures demonstrably have different standards of beauty, and within limits this can be acknowledged. However, as discussed in the previous chapter, a broadly generic aesthetic related to symmetry and obvious signs of health and absence of disease, e.g., a clear complexion, shows up universally. Normal human subjects have no trouble making discriminations based on such aesthetic standards. The consequences for this are far-reaching. People generally seen as attractive are ranked as more desirable mates, of course, but they are also more likely to be offered jobs and their personalities are rated more highly than those who are less attractive, among many other advantages (Rhode 2010).

So the fact that aesthetic beauty is undemocratically distributed in the population will have consequences for how people are ranked on some axes of status and resources. In chapter two, I reported briefly on the evidence that the second quality of status ranking Trump mentioned—intelligence—is also unevenly distributed in the population, even acknowledging that environmental factors play an important role in individual and group IQs. For reasons having to do with the nature of inclusive fitness, superordinates who have achieved their position for reasons of pure luck will have a compelling interest in maintaining advantages they have gained in the hopes of passing them along to their descendants, and absent intensive engineering efforts to prevent them from doing so, this will likely bear significant long-term consequences. For these reasons, hierarchies and stratified systems of status are quite a sticky feature of human society. Just as we saw that in the case of sex difference, the conventional logic—class-based domination by males of females with the end of cementing patriarchy—is less helpful for explaining the situation than a natural selection logic of real sex differences producing different distributions of attributes and attitudes by men and women that lead to a sexed division of labor, here too natural selection gives us a more robust view than the moralizing narrative of—in the Marxian terms that are often at work, at least implicitly—the bourgeoisie as a class usurping political power from the more numerous and morally righteous proletariat. It is perhaps less morally satisfying to understand that the class system, and other stratified hierarchies throughout human history, are a matter of the working out of a basic sociobiological template at the micro-individual level, over and over again in interactive contexts, but it is nonetheless more scientifically robust.

The basic behavioral mechanisms that enable stratification and status hierarchy are, as is typically the case, not unique to humans. Dominance seeking and aggression are ubiquitous among primates, and their connection to the emotion of anger is equally universal (Cabral and Almeida 2019). Formidability assessment mechanisms—that is, means by which judgments can be made about how likely another individual is to be willing and able to exert dominance over you—are widespread in the animal kingdom, and the evidence for their presence in humans is strong (Durkee, Goetz, and Lukaszewski 2018). Humans are attuned to a vast array of social cues that tell them about dominance hierarchies, their place in them, and the possibilities for rising in them. We have a predilection to associate lower voices with greater leadership ability, even though there is no clear empirical evidence to support that presupposition (Klofstad and Anderson 2018). We can and do make judgments about someone's place in a status hierarchy without verbal information, simply from visual cues such as how frequently the individual engages in expansive display (Tracy, Steckler, Rendles, Mercadante 2018). Human and some other primate stratification systems operate according to at least two different means for attaining hierarchy. Dominance produces hierarchy through the emotion of fear, whereas prestige signals expertise and skill, frequently mingled with a prosocial demeanor. There is evidence that human status systems based on dominance tend to be much more difficult to sustain over time than those based on prestige, but both mechanisms are ubiquitous in human history and society (Redhead et al. 2019). We would expect that here too the game of deception might play an important role, and the evidence suggests it does. Even in simple societies, there is evidence that systems of hierarchy among, for example, hunters are based not only on actual hunting skill, but also on deception as less than outstanding hunters dedicate significant effort, and with some success, to cultivating reputations as more skillful than they really are. The benefits of such status tend to outweigh the possible costs of being recognized as deceitful. It is not hard to find examples of such status-seeking gamesmanship in our modern world.

Humans have a curiously ambivalent relationship to status stratification systems. We are critical of the operation of such systems in the abstract, and especially when they negatively affect us personally, but we nonetheless avidly endeavor to climb in them and to enjoy the spoils of higher positions. Ultimately, our view of the subject is not so very unlike that revealed by the capuchin monkey experiment on inequality undertaken by Franz de Waal and discussed in his much viewed TED talk on the topic. The experiment ran as follows: You set up two capuchin monkeys in side-by-side cages, each easily viewable by the other, and have both perform a simple task—exchange one object for another, for example—and then reward them, but reward them unequally. When the task is completed successfully, you give the first monkey a grape, and the second monkey receives a cucumber slice. Capuchins like and will eat both, but—probably not unlike most humans—they greatly prefer grapes. Quickly, the monkey receiving the cucumber slices, seeing his comrade beside him getting a better payoff for the same work, becomes agitated. He throws the cucumber slice at the person who has given it to him, grabs the bars of the cage, rattles them furiously, and vocalizes his displeasure. Evidence of distaste for inequality? Well, yes, but of a rather distinctly limited kind. It is of note—though this part of the research is only infrequently reported in the media accounts of de Waal's findings—that only the monkey who receives the cucumber complains. The grape-receiving monkey, the one at the top of the unequal system of status and hierarchy, never expresses the slightest compunction about being unfairly rewarded and the plight of his unjustly deprived neighbor. Humans of course like to emphasize the fact that we at least occasionally do react to what we perceive as unfairness even if someone else is its target. But we probably do well not to overly exaggerate the difference. How many middle-class Westerners who hold to basically utilitarian values on material wealth disparities and who are aware of the level of suffering in poor parts of the world send any significant portion of their own resources to charities serving those sufferers? The figures indicate not nearly as many as you would anticipate if we were so very different from the capuchins.

Stratification systems can produce stress in their maintenance, especially when dominance rather than prestige is heavily weighted as the mechanism undergirding them. Robert Sapolsky's research on baboons is illustrative of a general situation in many primates. The species he studied are in a favorable ecosystem, and so they only need three hours of work a day to get their required calories. What do they do with the rest of their waking hours? Sapolsky (2017) reveals that a good deal of that time is taken up tormenting one another, superordinates moving around regularly to physically harass and intimidate subordinates and remind them of their place. Though stratification systems in modern societies do not typically involve the same kind of direct tormenting of subordinates by superordinates, they can certainly be stressful for those at the bottom rungs, especially in settings in which mass media and educational institutions make constantly apparent to them the material distance between themselves and those above them in the hierarchy. The historical lesson that no individual, however low ranking he might be, in a modern industrial society is nearly as materially poorly off as a peasant or expendable in an advanced agrarian society is a tough sell to the contemporary underclass youth with a smart phone that instantly and constantly shows him images of all the things the wealthy have and do that he does not and cannot.

Status Hierarchies and Sexual Selection

A great deal of status and stratification in humans emerges from the basic facts of reproductive competition, which we discussed in detail in the previous chapter. Sexual selection is a variety of selection that shapes organisms according to their success at getting access to mates. Intersexual selection is, in humans, typically about females choosing males with certain characteristics more frequently and thereby making it likely those characteristics will proliferate in males because those who have them will have more reproductive success. Intrasexual selection in our species is most pronounced among males. Darwin hypothesized that the greater size and aggressiveness of males replicated the pattern found in many other primates, in which males can often be observed engaging in threatening and physical competition amongst themselves for sexual access to females. His argument for an explanation was broadly consonant with what contemporary evolutionary scientists believe about the nature and consequences of sexual selection, which were discussed in the previous chapter. Anisogamy in our species means that in natural environments, i.e., those that are not significantly altered by cultural practices that endeavor systematically to undo these differential effects, male and female reproductive strategies can be expected to differentiate strongly. The optimal male strategy likely involves the effort to increase mating opportunities, which means increased competition with other males. This advantages those males who are larger, stronger, and more aggressive. It also advantages those who are prone to sexual jealousy. The evidence from preliterate peoples of significant levels of lethal male-on-male violence is significant (Pinker 2011). Napoleon Chagnon gives a compelling example of a conversation with a Yanomami tribesman who, after being told by Chagnon that some of his fellow anthropologists thought Yanomami violence was about conflicts over meat rather than women, responded "We like eating meat, but we like women much more!" (cited in Symons 1979:151). Indeed, there is evidence that both human and chimp males have used hunting prowess as a means of procuring sexual access from females in direct exchanges of meat for sexual services.

For the purposes of explaining status hierarchies and inequalities, perhaps the most relevant piece of evidence in human evolutionary history, though, is the way that sexual selection was shaped into a collective phenomenon in a species of significant strategizing intelligence. Lone powerful males in primate species might do reasonably well in the game of inclusive fitness, but males who were adept at ganging up with allies for the purposes of asserting status dominance might do even better. The rudiments of something we might reasonably call "politics" can be seen in the male chimp propensity to cultivate allies with whom later to plot and execute efforts to remove top males and take their places in the status hierarchy. Successful males in this framework are not necessarily the biggest and strongest, but rather those who are most skilled at planning and organizing what Richard Wrangham (2019) calls proactive aggression, and especially in groups. Human politics—even in recent times—gives ample evidence that leadership positions for males have very frequently been rewarded with greater sexual access to women. What feminists refer to as "the patriarchy" is an essentially indefensible concept insofar as males as a group clearly not only do not cooperate strategically against females as a group, but even fight viciously and lethally amongst themselves over and in the interests of females. Yet as I described in the previous chapter, the core outline of a human stratification system built on coalitions of males achieving dominance and status through explicit strategizing, motivated at base by the desire to achieve greater access to females, which might well include controlling them more effectively, is consistent with the theory of sexual selection. Nepotism, that is, kin selection, provides a strong explanation of the way in which such systems, once created, tend to be self-sustaining.

In contemporary democratic and largely meritocratic societies, however, it might be argued that none or little of this framework applies. Here, we actively try to undermine efforts by those who have attained high status and unequal power through anti-nepotism law and the cultural valuing of individual talent as the mechanism allowing access to high prestige and status. It is not so easy though to escape the logic of the theory of natural selection as a contributor to patterns of stratification. Assume a perfectly meritocratic society in which, say, objective testing was used as the means to determine who occupied the top rungs. We would still have the very considerable heritable aspects of IQ to deal with as contributors to outcomes. More, other personality traits that we know are related to the ability to do well in contexts of industrialized modernity—for example, discipline and self-control—are also highly heritable. As much as 60% of the variation in self-control might be genetic. If such heritability of personality traits conducive to achievement is combined with a growing tendency of elite groups to more or less exclusively pair and reproduce among themselves (see e.g. Murray 2012), it is not implausible to imagine the emergence of a fairly durable system of inequality based on these patterns, in which those at the bottom are highly likely to stay there even in social orders with the democratic, egalitarian ethos of the modern cultural world.

The evidence that hierarchy is an inevitable and, in many ways, a productive feature of human societies only begins with the fact that we know of no societies that do not stratify. We have also the evidence that, in the modern human societies that have been most theoretically committed to eliminating hierarchy and status—the experiments in communism—the results were catastrophic in terms of the broad state of those societies, and hierarchy merely transformed itself into new forms. In the modern democratic societies, where liberal individualism as a political doctrine has become dominant, the variety of status system is theoretically meritocratic, based on individual talent, but that is inevitably compromised by the facts given in this chapter concerning human nature. More radical political efforts to undo hierarchy typically engage in practices of affirmative action and preference for members of groups seen as disadvantaged, but they give little more evidence of success at undoing status systems. As in the communist efforts, they simply create new hierarchies that demonstrate a symbolic valence that is merely a reversal of the system they desired to replace. Coming to terms with hierarchy may be the single most important step required for a mature human science.

Note

1 The summary that follows of inequality from hunter–gatherer to industrial periods is largely drawn from Nolan and Lenski (2015) and Lenski (2005).

References

Cabral, João Carlos Centurion and Rosa Maria Martins de Almeida. "Effects of Anger on Dominance-seeking and Aggressive Behaviors." *Evolution and Human Behavior*, 40 (2019): 23–33.

Durkee, Patrick K., Aaron T. Goetz, and Aaron W. Lukaszewski. "Formidability Assessment Mechanisms: Examining their Speed and Automaticity." *Evolution and Human Behavior*, 39 (2018):170–178.

Flannery, Kent and Joyce Marcus. *The Creation of Inequality: How Our Prehistoric Ancestors Set the Stage for Monarchy, Slavery, and Empire*. Cambridge: Harvard University Press, 2012.

Gaxotte, Pierre. *The French Revolution*. Translated by Walter Alison Phillips. New York: Charles Scribner's Sons, 1932.

Kelly, Robert. *The Lifeways of Hunter-Gatherers: The Foraging Spectrum*. New York: Cambridge University Press, 2013.

Klofstad, Casey A., and Rindy C. Anderson. "Voice Pitch Predicts Electability, But does not Signal Leadership Ability." *Evolution and Human Behavior*, 39 (2018):349–354.

Lenski, Gerhard. *Ecological-evolutionary Theory: Principles and Applications*. New York: Routledge, 2005.

Murray, Charles. *Coming Apart: The State of White America, 1960–2010*. New York: Crown Forum, 2012.

Nolan, Patrick and Gerhard Lenski. *Human Societies: An Introduction to Macrosociology*, 10th edn. New York: Oxford University Press, 2015.

Piff, Paul, Daniel M. Stancato, Stéphane Côté et al. "Higher Social Class Predicts Increased Unethical Behavior." *PNAS*, 109, 11, (March 13, 2012):4086–4091.

Pinker, Steven. *The Better Angels of our Nature*. New York, NY: Viking, 2011.

Redhead, Daniel, Joey T. Cheng, Charles Driver, Tom Foulsham, and Rick O'Gorman. "On the Dynamics of Social Hierarchy: A Longitudinal Investigation of the Rise and Fall of Prestige, Dominance, and Social Rank in Naturalistic Task Groups." *Evolution and Human Behavior*, 40 (2019):222–234.

Rhode, Deborah. *The Beauty Bias: The Injustice of Appearance in Life and Law*. New York: Oxford University Press, 2010.

Rothman, Joshua. "The Equality Conundrum: We All Agree that Inequality is Bad. But What kind of Equality is Good?" *New Yorker*, January 13, 2020.

Sapolsky, Robert. *Behave: The Biology of Humans at our Best and Worst*. New York: Penguin, 2017.

Schwalbe, Michael. *Rigging the Game: How Inequality is Reproduced in Everyday Life*. New York: Oxford University Press, 2007.

Symons, Donald. *The Evolution of Human Sexuality*. New York: Oxford University Press, 1979.

Tocqueville, Alexis de. *The Old Regime and the French Revolution*. Translated by Stuart Gilbert. New York: Anchor Books, 1955.

Tracy, Jessica L., Conor M. Steckler, Daniel Randles, and Eric Mercadante. "The Financial Cost of Status Signaling: Expansive Postural Displays are Associated with a Reduction in the Receipt of Altruistic Donations." *Evolution and Human Behavior*, 39 (2018):520–528.

Trautmann, Stefan, Gijs van de Kuilen, and Richard J. Zeckhauser. "Social Class and (Un)Ethical Behavior: A Framework, with Evidence from a Large Population Sample." *Perspectives on Psychological Science*, 8, 5 (2013):487–497.

Wrangham, Richard. *The Goodness Paradox: The Strange Relationship between Virtue and Violence in Human Evolution*. New York: Pantheon, 2019.

7 Racial Identity and Difference in *Homo sapiens*

In 2012, an essay appeared in *Sociological Theory* (Shiao et al. 2012, referred to hereafter as "Shiao") that proffered a way of theorizing race in the social sciences as a response to what the authors described as the inability of the dominant social construction of race approach to adequately account for the phenomenon. For many sociologists, the approach presented in the article probably appeared strikingly novel, but it was deeply rooted in established scientific ways of thinking about how genetic and phenotypic variation work across animal populations of the same species living continuously in large territories with different ecologies and with a significant degree of breeding isolation from one another for long periods of time. The authors, in a spirit of generosity to those whose work they were challenging, acknowledged that the term "race" might be too marred by its sometimes sordid history to rehabilitate and instead offered another term, "clinal class," to refer to ancestral population groups with such differing evolutionary histories.

The response to the article was informative regarding the sternness of the refusal in the contemporary discipline to take seriously the idea of thinking about human genetic diversity scientifically. Two years after the article was published, the journal ran a special symposium issue full of dismissive responses to the Shiao article. Every one of them rejected outright any scientific utility to the approach delineated, and several suggested the framework might be racist, or at least might easily make itself available as an intellectual tool to racists. There was no suggestion in any of the responses of the promise of the approach offered, however introductory and limited it might be. The responses merely reasserted the superiority of the mainstream social construction of race approach—often while demonstrating serious misunderstanding or mischaracterization of the argument in the original article—and claimed that the fact that the boundaries separating human racial groups are fluid and are influenced by the decisions of the analyst scientifically invalidates meaningful and scientifically useful efforts at categorization of the human species into racial subgroups. One of the responses even included as co-author the biologist Richard Lewontin, perhaps the single figure in biology most cited by adherents to the social construction of race perspective due to an article he produced in the early 1970s on genetic variation in our species, in order to bolster the credibility of their wildly inaccurate accusation that the Shiao article constituted little more than "a latter day version of eighteenth-century attempts to biologically classify humans into racial categories" (Fujimura et al. 2014:219).

After this uniform and stern denunciation by a unified group of adherents to a competing theoretical perspective, the article and its suggestion for carefully but rigorously intellectually opening the sociology of race to evolutionary and genetic research sank like a stone. I have never seen it cited subsequently. Neither the main author nor any of the co-authors of the piece has published on this topic since, to the best of my knowledge. When I reached out to the main author to inquire about any subsequent efforts to follow up on this initial effort, he told me his work in this area had largely been paused since the publication of the symposium issue.

The Biological Reality of Human Genetic Diversity and the Feasibility of Referring to that Diversity with the Term "Race"

What is race, at bottom, for a social scientist? Briefly put, it is a consequence of the reality of human population genetic diversity and the natural selection mechanism of attraction to in-groups and aversion to out-groups. But the mainstream of the discipline of sociology in 2020 denies the existence of human population differences that could reasonably be defined in this way. Sometimes this consists of a denial of the categories themselves, which are described as social constructions, while still acknowledging real biological differences in populations. But some mainstream texts in the discipline deny the reality of significant genetic differences in regional populations altogether or deny that such differences could conceivably have any possible effect on anything that social science considers important. Here is an example of how bluntly one commonly used textbook defines the term: "Race is a modern social construction, meaning that the idea of race is not based on biological differences among people, even though race has become important in determining how we interact. It is a particular way of viewing human difference that is a product of colonial encounters" (Golash-Boza 2015:6). This and many other introductory texts on the topic in the social sciences largely ignore and leap over the biological science of human population difference. They will often simply repeat variations on the same theme, i.e., that members of our species share more than 99% of our genomes and therefore any differences at that genetic level by definition cannot be anything but minimal and superficial. Sometimes, no mention is made of any genetic contribution to human difference at all. The text just cited, for example, has only one reference in its index to genetics, and this is to an exceedingly brief passage in the book attacking the "myth" of diseases that are unique to specific racial groups.

It is of course true that all members of our species are genetically quite similar. This similarity is part of the definition of the term "species." But the questions that are not asked by the mainstream texts are precisely what genetic variation in the species looks like and what consequences it might have on human behavior and social organization. To a fair degree, the claim "all humans are 99+% the same" relies for its rhetorical power on a basic ignorance of the sheer size of the genome and the statistically small amount of genetic difference required to produce profound phenotypical difference given that size. The total number of base pairs in the human genome is around 6.4 billion. The number is sometimes given as 3.2 billion, but this derives from the reference genome compiled by the Human Genome Project, which gave only the DNA sequence for single copies of the 22 non-sex chromosomes, the X and Y sex chromosomes, and the mitochondrial DNA. The actual amount of DNA in a human being is roughly twice that because our genome is diploid, that is, we have a pair of each of the chromosomes, one from our mother, one from our father. Given the size of the genome, even a miniscule fraction of variance produces many real genetic differences. Evidence suggests that the typical human individual's genome differs from the reference genome of the Human Genome Project at around 5 million sites (The 1000 Genomes Project Consortium 2015). That is more than enough genetic variation to generate real phenotypical difference, depending on which base pairs in which genes are in play in the variation. This is an empirical question that, in the human genome, we are only just beginning to unpack and address. Another way to demonstrate the ignorance relied on in the insinuation is to show how much genetic similarity exists between two species that are clearly phenotypically different. Chimps and humans, for example, share just under 99% of their genomes. It is a mere 1% difference in their genomes that generates the massive, undeniable distinctions we see between us and our closest primate relative in appearance, intelligence, and behavior.

Those who would deny any meaningful scientific utility of the category "race" will often point out that there has been significant disagreement among scientists endeavoring to define how many races exist. Darwin himself pointed to this diversity of opinion, while yet affirming the scientific utility of racial difference as a concept. It is true that there has been considerable disagreement on the precise number of human races. But this does not invalidate the idea that human groups can be reasonably differentiated based on characteristics derived from the combination of different genes and exposure to different environments, or that such differentiation and categorization might not only be something scientists do but that the groups themselves might with some accuracy use those real genetic differences as a basis for thinking about identities and acting in the world.

The reason that it is of no shattering importance that one theorist has argued for five races and another for fifteen or even fifty is that this is an inevitable situation given the way variation in the human population works. Human population variation is continuous, which means that what we think of as race is a variable that exists on a continuum, rather than as a set of naturally existing discrete categories. In the Shaio article with which I started the chapter, it is pointed out that the classic social constructionist account of race, like the biophobic accounts in other realms we have already discussed, situates itself in a complex discourse by a simplifying move that is not justified by the evidence. This perspective, represented in the article by a book (Omi and Winant 1986) that has become one of the classic sources for constructionists, argues that race has no biological content at all, and that it is entirely and only a conceptual grid produced by "social, economic, and political forces" (Shiao et al. 2012:67). Such forces do indeed have something to contribute to the substance of racial categories. It is not hard to see, for example, how much the substantive content—especially at the level of connotation and emotional meaning—of the racial categories "black" and "white" in nineteenth-century America was influenced by an existing system of stratification based on a limited set of phenotypical features. But this does not invalidate the significant evidence in support of real biological content that differentiates human groups. Gradual variation in phenotypical traits, which we can now demonstrate to be tied in important ways to variation at the genetic level, is undeniably present in the human global population. One finds it hard to see such variation from ground-level, but imagine the following thought experiment: You are flying above the surface of the planet with a viewing device that allows you to see human beings both in their broad phenotypical characteristics (skin color, body morphology and size, facial features, eye shape, hair color and texture, etc.), but also at a distance that permits a scope of vision that encompasses large populations in the way they are often represented graphically on world maps. You would see, if you started, say, at the southern tip of the continent of Africa and then flew northward over the continent, and then up through Morocco and into Spain, then northward through Europe to Scandinavia, a gradual shift in those population phenotypes that would, over sufficient distances, appear quite starkly obvious and undeniable. But the change would happen subtly and incrementally, not categorically, all at once. That is, at no point would one be able to draw a line and say 'Objectively, this is the one clear point in the continuum at which racial group A ends and racial group B begins.' You could easily move the line a little bit northward, or a little bit to the south and then make the same statement, and it would be equally false. But it would also contain the truth that change, phenotypically, was clearly happening. And we know now that this change is genomic as well.

The critics of the idea of racial groups are correct that no such groups are given as discrete, categorical distinctions in the natural world. They require human beings—whether scientists or just ordinary members of social groups—to observe and act on them. But this does not mean there is no real population diversity on which the scientists and non-scientists can make their distinctions. And especially when we are talking about scientific efforts to describe human phenomena, racial or clinal class categories might well be useful notwithstanding the fact that different scientists with different intellectual tasks might find a different set of racial or clinal class groups useful. So long as they are based in the actual variation in the population, all such categories can be correct. It is simply a matter of the level of relief at which you concentrate your analytic attention. The racial differences between two sub-groups in a population can be genetically demonstrated, for example, by showing a difference in a specific number X of alleles and a consistency of genome within groups of some specific percentage Y, then when you look more closely, you find, for example, each of those two groups can be cut into two smaller groups, which differ from one another on some number of alleles less than X, and a consistency of genome within groups of something greater than Y%. The original categorical schema of two groups is scientifically correct, and so is the second schema of four groups. The evolutionary biologist Theodosius Dobzhansky, in a book on the evolution of the human species that ought to be read by every social scientist, made the key point about as succinctly and clearly as it can be made: "[I]t should always be kept in mind that while race differences are objectively ascertainable facts, the number of races we choose to recognize is a matter of convenience" (1962:266).

The classificatory principle here is, it should be emphasized, the same as that used to define social class groups. Where is the objective line between the working class and the lower middle class? It does not exist in nature. We social scientists decide where the line demarcating the two categories will exist for the purposes of any given study—or, as is more common, we just omit this careful step altogether and nonetheless continue to use the category terms in a loose, unscientific way. Is the income level separating one class group from another $40,000/year or $50,000/year? It depends. Careful sociologists will feel the intellectual need to make some defense of their particular definition of the categories, but they make the categories nonetheless. Tellingly, no sociologists I have ever encountered raise any objections whatever to the use of this classificatory strategy on class along the lines of those we hear constantly in the discussion of race, nor do any I know accuse those who insist on making such categorical distinctions of social class of engaging in the harmful reification of the social constructs of material income and wealth difference. It is universally recognized, at least by sociologists who have thought carefully about this topic, that class categories are, on the one hand, malleable and constructed by scientists and, on the other, based in real features of variation in the relevant population at the same time. Rigorous categories of race meet the same criteria.

All serious study of the human genome shows that there is genetic variation in our species and that this variation can be seen both between individuals—even those who are closely related—and between members of ancestral population groups with different evolutionary histories. In the first systematic effort in the era of modern genetics to describe the genetic geography of our species, Luigi Luca Cavalli-Sforza, Paolo Menozzi, and Alberto Piazza recognized the great difficulties involved in accurately accounting for the regional differences, and they rejected the term "race" because of its history of entanglement with "prejudice, misunderstanding, and social problems" (1994:19). However, they then went on to show that a genetic linkage tree can be drawn using the 42 populations they studied on 120 gene allele frequencies that maps remarkably well on to the conventional set of racial categories most readers of the book would have known and at least informally used in making sense of human identities before reading the book. That is, there emerged from their analysis three broad human population subgroups—African, North Eurasian, and Southeast Asian. They then grouped the 42 populations into nine clusters based on similarity of genome: 1) sub-Saharan Africans, 2) European Caucasoids, 3) extra-European Caucasoids, 4) Northern Mongoloids, 5) Northeast Asian Arctic populations, 6) Southern Mongoloids, 7) New Guineans plus Australians, 8) those from minor Pacific Islands, and 9) Americans (ibid.:78–80). In their words, "[t]he single most important conclusion…is that the greatest difference within the human species is between Africans and non-Africans" (ibid.:83). The magnitude of the genetic difference between these two groups was more than two times greater than that between the Southeast Asian and North Eurasian groups (ibid.:80). More recent large-scale analyses that have examined still larger numbers of alleles show quite similar groupings.

Dobzhansky provided a comparative angle from non-human genetics on why racial classification in humans was potentially useful and yet subject to the kind of categorization variation just noted:

> One should distinguish two aspects of the problem of "reality" of race—race as a category of classification and racial differences as data of observation … We have seen that populations of *Drosophila pseudoobscura* in parts of the United States and Mexico and even at high and low elevations of the Sierra Nevada range differ in their relative frequencies of the gene arrangements in their third chromosomes. These are racial differences. We have also seen that human populations of different countries have diverse frequencies of certain blood groups. These are likewise racial differences. There is nothing arbitrary or unreal about these differences. However, classifying races or subspecies is an altogether separate issue. One could affix racial names, in English or Latin, to populations of Drosophila pseudoobscura that inhabit the Pacific Coast of the United States, the Rocky Mountains, and northern and central Mexico. But this would hardly serve a useful purpose; saying that a given population is of such-and-such geographic origin is good enough. To most students of human racial variation racial labels seem convenient … The number of names that one wishes to have is again a matter largely of convenience.
>
> (1977:144–5)

He then suggest a schema, strikingly similar to the one just described in Cavalli-Sforza, that is provided by Stanley Garn (1961) with a "proposed… subdivision of the human species into nine 'geographical races' or 34 'local races.'" The nine geographical races are Amerindian, Polynesian, Micronesian, Melanesian-Papuan, Australian, Asiatic, Indian, European, and African (ibid.:145–146).

Dobzhansky defended the scientific rigor of the concept of race in the human species against one of the leading lights in the movement in modern cultural anthropology to undermine it, Ashley Montagu. *Homo sapiens*, he argued, is a "polytypic species," that is, a species with significantly differentiated sub-groups that can scientifically be classed as genetically distinct. The term "race" as it is used in humans is in fact a synonym for subspecies, and racial groups in humans are equivalent to e.g., the subspecies variants of the golden whistler, a bird of the Solomon Islands. The definition of the term "race" Dobzhansky gives is "Mendelian populations or arrays of genotypes that inhabit parts of the distribution area of a polytypic species" (ibid.:137). Elsewhere, still more succinctly, he gives this definition: "subordinate Mendelian populations within a species" (1973:57). A Mendelian population is simply a group of organisms that interbreed and share a common gene pool. Racial differences in nature have a range from minimal, having to do with frequencies of just a few genes, all the way to distinctions between populations in sufficient reproductive isolation to teeter on separation into different species (Dobzhansky et al. 1977:137).

It is often claimed by opponents of this approach to race that the number of individuals who elude the categories—those who are racially hybrid, or mixed—is so numerous as to disrupt any classificatory utility. A typical pedagogical example advocates of this idea utilize to attempt to undermine the race concept is to display images of individuals who are phenotypically interstitial to existing racial groups, or who might reasonably be placed visually in two or more categories. The reality is that the self-identified racial group and the objective grouping data of one's genome are very nearly perfectly correlated in actual studies of the relationship. In one study of the relationship between these two definitions of race, barely one-tenth of 1% of subjects showed genetic cluster membership that differed from their self-identified racial identity (Tang et al. 2005). The subjects, from a range of different locales in the US and in Taiwan, of this study were asked to identify themselves as a member of one of four major racial/ethnic groups—white, African-American, East Asian, and Hispanic. The genetic analysis of the subjects also revealed four main groups or clusters and overlap between the identities and the genetic analysis was nearly perfect. Only five of the 3,636 subjects did not show correspondence.

Dobzhansky died in the mid-1970s, and contemporary critics of his vision of the biological basis of human ancestral populations frequently claim that the scientific consensus today opposes his position, which they dismiss as antiquated. But there are contemporary biologists and geneticists of considerable profile in the scientific community who agree with Dobzhansky's general perspective on this topic. Jerry Coyne, an emeritus professor of biology at the University of Chicago, has articulated a similar view of race or sub-species in humans on his widely read blog (https://whyevolutionistrue.com/2012/02/28/are-there-human-races/). Even more notably, the Harvard geneticist David Reich has, in both a recent book (Reich 2018) and a *New York Times* op-ed that accompanied its 2018 publication, argued firmly for the existence of genetic differences among human population groups with different ancestral profiles and for the need for continued serious research into the nature and extent of such differences. He notes that the now dominant orthodoxy in many academic circles that genetic differences between human ancestral populations are so minimal as to have no possible implications for cognitive and behavioral traits is simply indefensible in light of current knowledge. Genetic data alone, for example, gives us a "far from trivial" ability to predict an individual's years of education, and there is no good reason to suspect that further research into group differences on this and other such variables can only reveal trivial such differences (ibid.:256, 258). Reich makes it clear that, while there is a risk that genuine racists will potentially find comfort in some of what we are learning about the genetics of human difference, the risk of retreating from scientific inquiry for political reasons and pretending no such data exist is far greater, as it cedes the field of discussion of the genetic discoveries to those without scientific expertise.

While social scientists who deny the utility of the notion of race often evince suspicion at scientific evidence for it, frequently on the grounds that the science is purportedly too mired in the racism of human society, they tend to trust blindly in other items of scientific consensus that are in fact contested. For example, they will often defend their rejection of the potential importance of genetic diversity in our species by reiterating that, whatever our superficial differences, "all people who were ever born anywhere on Earth have been, are, and will be descendants of Africans" (Krimsky and Sloan 2011:31). It is indeed the current conventional understanding that modern *Homo sapiens* evolved once, in Africa, and that all contemporary humans directly descend from this original group. But there is a significant contender to the out of Africa theory that has adherents in the scientific community, a formidable logic, and data to support it. The multiregional hypothesis on human origins argues that *Homo erectus*, a predecessor of our species, originated in Africa, but some members of this species then migrated out of Africa and that subgroup evolved independently a number of times into the several regional variations of modern humans. A major variant of the multiregional hypothesis posits that regional variants of *Homo erectus* in Africa, Europe, Asia, and Australia gave rise to the modern populations of those four groups. There is, some scientists attest, sufficiently significant genetic variation among the four to make the theory plausible (Lee 2015:223–224; Coyne 2009:206).

The lack of sharp boundaries between the subordinate Mendelian populations that make up racial groups in humans can be frustrating for many. But this point has nothing to do with its scientific validity, and we must follow the science where it goes. A political philosophical commitment to human equality protects us as much as possible from dangers, but the truth on this cannot simply be abandoned on the shaky argumentative grounds of political correctness. For "diversity is an observable fact of nature, while equality is an ethical commandment" (Dobzhansky 1973:4). Nothing in the science of human difference can legitimately be used as justification for unequal treatment of citizens under law, and we can take comfort in this distinction between scientific and moral/legal discourses.

And yet the situation in sociology today is decidedly inhospitable to examining the question as one of the leading biological scientists of the twentieth century suggested it should be examined. Even more genetically-informed sociologists who intervene on the question of the viability of race as a concept tend to skew toward the ideological consensus that it is a mere social construction with no basis in biological science. Dalton Conley and Jason Fletcher are exemplary here. In a recent book on genomic sociology, they claim that because there is greater genetic distance between two African groups (the Fulani and the Hadza) than between East Asians and Europeans, the reality and the efficacy of the concept of race is disproven (2017:87). But this logic is easily seen through by reference to the figure to which they are referring, which they reproduce from a 2009 *Science* article (Tishkoff et al. 2009). This figure graphically illustrates the proximity of genetic relationships between groups by the length of the line connecting the groups. They have chosen for their demonstration a tiny group, the Hadza, that is a genetic outlier compared to every other regional group on Earth. When one compares e.g., the distances separating all non-Hadza African groups—which are uniformly small—to the much greater distances between the Africans as a cluster and every other global population, or the small distances separating the different African groups to the much greater distance separating East Asians and Native Americans, or the distance between either of these last two groups and Europeans and Indians, five racial group categories—African, East Asian, Native American, European (including Indians), and Oceanian—that are quite compatible with the traditional such categories emerge clearly as discrete entities.

Conley and Fletcher further give away their conventional sociological biases on this topic when they examine the complicated topic of racial/regional populational differences and behavioral predilections and personality traits. They leap over complicated discussions in the literature on group differences on IQ by gesturing at the stock of answers traditional sociologists give on this question, which point always as a default at racial discrimination and stereotyping as the entirety of the explanation of such differences. In an appendix to the chapter, they uncritically reiterate a misrepresentation of research on stereotype threat as the cause of black/white differences in test performance that has become one of the most common errors repeated over and again in mainstream sociological literature on race (Sackett, Hardison, and Cullen 2004). They are certainly correct to reject the claim that science has fully answered the diverse, complex, and controversial set of questions it is currently pursuing regarding the causes of empirically demonstrable racial group differences in a range of important behavioral traits and social outcomes. Those questions have not yet been answered, and it is premature and unscientific to claim that we know those answers. The real question though is whether we will have the intellectual courage to continue to seek scientific solutions to such puzzles. Though Conley and Fletcher end their chapter on race admirably noting "we cannot be afraid to look" at the topic, the substance of the chapter mostly avoids the most interesting contemporary biological science on human diversity (2017:112).

With some frequency, social scientists who wish to dismiss the biological side of the categories of population difference will claim the biological science is on their side. Almost inevitably, they will invoke one or the other, or both, of two figures: Stephen Jay Gould and Richard Lewontin. These two were the central figures involved in the concerted effort to undermine E.O. Wilson's efforts in the mid-'70s to turn the social sciences in the direction of sociobiology and evolutionary theory. When Wilson's *Sociobiology* was published in 1975, Gould and Lewontin, who were both at Harvard along with Wilson, mounted an energetic counteroffensive that was often transparently motivated more by cultural politics than by science, as I discussed in chapter two. In 1981, Gould published a book, *The Mismeasure of Man*, that essentially labelled all attempts to analyze race by physiological, biological data as racist. A central element of the book's argument was a denunciation of the work of the nineteenth-century biologist Samuel Morton to classify racial groups by skull measurement. According to Gould, Morton mismeasured the skulls with the deliberate intent of skewing data to fit his own racist prejudices. However, retesting of the skulls found that while Morton had made errors, they were not distributed in such a way as to indicate malicious intent. In fact, he had overestimated the cranial size of many black skulls, which would have undermined any effort to fudge data in a racist manner. It turned out that it was Gould who had done the substandard analysis of Morton's data and apparently falsified with an *a priori* agenda (Lewis et al. 2011).

Lewontin had, by the time of the publication of Wilson's book, already produced the 1972 article earlier referenced that purported to show that racial group classification had no genetic basis. The argument here was that far more genetic variance—nearly nine-tenths of the total—was apparent in comparisons of individuals in the same racial group than in comparisons between such groups. A special 2001 edition of *Nature* on the human genome included a CD that reiterated this claim, and many standard social science texts cite this tidbit—often as the only reference to a biological scientist—in dismissing the biological category of race. But Lewontin himself admitted that his analysis relied on analysis of only a tiny number of genes—fewer than 20—and in fact, some of those he looked at showed significant between-race variation. The methodological framework of Lewontin's argument was demolished by A.W.F. Edwards (2003), who showed that the common interpretation of Lewontin's data—that no reasonable logic from genetics could undergird racial group classifications—followed from the fact that the analysis had considered only single genetic loci, rather than clusters of alleles at different loci. Edwards demonstrated that the racial categories Lewontin had claimed there was no evidence for are in fact robustly present once one moves from the "statistical fallacy" of Lewontin's reliance on single locus-to-locus comparison to the more appropriate analysis of correlations among different loci, that is, comparison of correlated sets of genes between different organisms. Edwards even argued that he and Cavalli-Sforza had suggested the superiority of this method for classifying human population groups at a conference Lewontin attended nearly a decade before his paper was published, but Lewontin failed to mention it altogether in the 1972 paper. The conclusion from Edwards as to the main claims made by Lewontin, and subsequently nearly endlessly reiterated in the conventional social science world, was clear:

> It is not true that "racial classification is ... of virtually no genetic or taxonomic significance." It is not true, as *Nature* claimed, that "two random individuals from any one group are almost as different as any two random individuals from the entire world," and it is not true, as the *New Scientist* claimed, that "two individuals are different because they are individuals, not because they belong to different races" and that "you can't predict someone's race by their genes." Such statements might only be true if all the characters studied were independent, which they are not.
>
> (ibid.:801)

It is an interesting commentary on the intellectual sphere in Anglophone societies that we appear to be outliers on the question of how to think about the concepts of race and racism. There is evidence that a persistent adherence to the mainstream social scientific view of race and an aversion for considering the scientific evidence of its reality seem to be particularly widespread in the culture of Western and especially Anglophone intellectual worlds and that they are much less pronounced, or even altogether absent, elsewhere. English speaking countries show the highest level of inattention to race as a social science category, while European countries show moderate inattention, and Chinese and Russian scientists more or less fully accept the validity of the category of race as rooted significantly in genetics and biology (Lieberman et al. 2004).

The Connection to In-Group/Out-Group Hostility and Aggression

It is difficult to argue coherently that there is no morally appropriate utility in knowing about at least some genetic differences between racial groups. As Cavalli-Sforza puts it, the "chances of having the same diseases, and responding similarly to the same drugs" are significant in genetically similar populations, and so a medical science that would most appropriately treat patients of different population groups must embrace such knowledge, rather than pretend it does not exist (2000:31). Once we have admitted that medical facts of importance will vary according to differences in human population genetics, we are face-to-face with the dilemma I discussed in chapter two with respect to my colleague who wants to keep the medical, the psychological, and the behavioral neatly separated. For in realms outside the medical, it is clear that certain physical traits and concomitant physical abilities vary along these group lines. For example, some athletic achievements that distinguish different population groups, such as the nearly total domination of long distance running by Kenyan runners of a single tiny tribe, the Kalenjin, representing 0.06% of the global population, are obvious to any non-casual observer.

It is true that none of the preceding regarding the biological bases of differences in races, or subspecies, or regional or ancestral populations, or clinal classes, or whatever category term we decide to give them would be meaningful to the social sciences if such differences had no relationship at all to human personality, behavior, and social organization. Merely posing that possibility raises suspicions among many in an atmosphere, such as that in most Western cultures, wherein a deep-seeded desire to see all humans as essentially equal—not just in political philosophical but in all empirical terms–exists. Add to this the more or less systematic effort by large segments of the academic world to deny such differences, and even to apply pressure of various kinds to those in their ranks who show signs of questioning the orthodoxy, and the question gets still more complex. There is nonetheless ample evidence that the relationship exists, and this is why sociologists must be able to effectively theorize race, racial identity, and how they affect social life.

We know that there is a kind of classification that human groups make easily and indeed readily, even enthusiastically. I talk about it with students as the "In-Group/Out-Group" or "Us/Them" game. The evidence for our facility at organizing the world and then acting on it in this way is abundant. Lopreato describes two behavioral tendencies visible in a wide range of human action. Homologous affiliation is "a genetically based inclination to favor alliances and forms of reciprocal behavior with others in direct proportion to one's degree of kinship and/or phenotypical similarity to these" (1984:304). Heterologous contraposition is the "biologically based predisposition to have negative attitudes toward others on the basis of real or perceived differences that are typically, though not exclusively, of an ethnic nature" (ibid.:332).

The logic here is intimately consistent with the remainder of what the theory of natural selection tells us about human interests and moral behavior, which I discussed in chapter four. Human interests are reproductive, which means they must be calculated according to inclusive fitness. Individuals can, all other things being equal, be expected to aid kin more readily than non-kin. It would have been the case that in early human societies of relatively small numbers, the members of a given society would have been more closely genetically related to one another than to most of the members of other social groups, even those with whom they intermarried to combat the dangers of incestuous overbreeding. It is not hard to hypothesize how modern racial group identities and antagonisms could have developed out of a history and a set of evolutionary forces that look like this. One need not even postulate any greater real familial relationship between, say, the average member of modern racial group A and other members of that group, and between that member of group A with the average member of another racial group B. In any empirical situation, it may well be that the individual in question is equally distant from both, or more closely related genetically to the member of the other group. The forces driving individuals to operate under the rules of kin selection and nepotism, and concomitantly to feel distance from non-kin, were not capable of giving individuals foolproof mechanisms to distinguish close kin from others. But among the more or less reliable mechanisms for doing so were almost certainly some of the phenotypical and ideational methods modern humans use to distinguish among different racial groups. Stereotyping certainly plays a role here too, and Steven Pinker (2002) has courageously discussed why stereotyping, whatever one thinks about its implications and its moral value in modern cultural spheres, has almost certainly been selected for in our species as a fairly reliable way of making judgments about, among other things, what to expect from interactions with others, especially others one does not know personally. I return to this topic at the end of this chapter.

It is not exceedingly difficult to demonstrate why ethnocentrism, driven to some degree by observable evidence of real genetic difference in populations, would have been a universal mode of identity formation in deep evolutionary time, something that would have emerged early in the life of our species, or in fact more likely, as we have seen, something we inherited from earlier predecessors. Computer simulations demonstrate that ethnocentrism as a strategy outcompetes a number of other strategies for interaction and reciprocity, including humanitarianism, and that it was almost certainly selected for in the evolutionary contexts in which our species existed for most of our time on the planet (Hartshorn, Kaznatcheev, and Shulz 2013). The adaptive qualities of xenophobia have also been tested in experimental settings and revealed to yield groups that outcompete groups that do not display xenophobia (McEvoy 2002).

Robert Sapolsky admirably summarizes much of the science on how trust and cooperation with "us" and aggression, or at least antipathy, toward "them" operates in the human brain, and in the brains of our primate relatives. Oxytocin, that marvelous neuropeptide discussed in earlier chapters, contributes significantly to these predispositions. There is even evidence that the innate—or at least very early origin—bias toward members of groups phenotypically different from self—about which the implicit association/bias test phenomenon has revealed much—operates in at least some non-human primates. Monkeys apparently associate other monkeys of their in-group with positive things, such as tropical fruit, and members of out-groups with negative objects, such as spiders (2017:389).

Yet, amidst the piles of data he discusses that confirm the phenomenon, Sapolsky's own moral and political commitments frequently lead him to make more of the counterevidence than is merited. He writes for example about a particular troop of baboons he was studying in the mid-'80s that had an accident fall upon it, inadvertently setting up an interesting experiment in the nature of aggression. There was a tourist lodge near the troop, and it established a garbage pit in which was deposited all the scrap food left by the tourists. Another troop of baboons, different from those Sapolsky was studying, discovered the garbage pit and began making daily visits, establishing it as their protected feeding ground. The baboons in Sapolsky's troop also found out about the pit, but the only members of that troop who dared to visit it and face the opposition of the members of the other troop were the most aggressive, physically imposing, least socially-oriented males. A tuberculosis outbreak emerged among the garbage dump troop, and this disease, which spreads very easily among baboons and is typically fatal in relatively short order, spread also to the males from Sapolsky's troop that frequented the site. Enough males from the troop died to create a large sex imbalance in favor of females. Sapolsky discovered that, over time, the normally extremely aggressive behavior of male baboons changed in this troop. The remaining baboons were, he assumed, the least aggressive in the troop, and they were now outnumbered 2:1 by the females. What happened was an astonishing transformation of culture in the group with respect to aggression. It plummeted. Males in the group even groomed one another, a practice basically unheard of in typical hyper-competitive male baboon culture. Sapolsky noted that the culture seemed relatively stable, as it was still present more than two decades later, and new male chimps who came into the group from outside seemed to adapt to the native culture.

How likely, though, is it that this is sustainable over the long haul, especially if there is any significant competition from other baboon troops or migration of outside males? In the published record, I could not find Sapolsky speaking much to this, but in an account of a lecture he gave on the topic that can be found online, one of the important aspects of the maintenance of the group's peaceful culture became apparent. It seemed Sapolsky had seen at least one instance of an attempt by an aggressive outside male to come into the group with the traditional aggressive behavioral package unchanged. The response on the part of the "peaceful" males in the group was to rip him limb from limb and to continue their mutual grooming while the outsider's detached head was sitting on the ground before them.

[1] In other words, not only were the baboons immediately able to invoke the aggressive preset of males of their species, their capacity for in-group solidarity and aggression and violence against out-groups was undiminished. Presented with a relevant context of Us/Them confrontation and competition, they responded with the behavioral program that evolutionary history has provided them. Short of enough deep evolutionary time in the kind of idiosyncratic situation created by these unique circumstances to significantly alter the genomic consequences of sexual selection, it is quite likely that this odd cultural group will not persist and will not export its culture outward. None of the physiological/biological aspects of sex difference in baboons has disappeared, and once the demographic balance is closer to 1:1, it is unlikely, especially in the context of a surrounding ecology full of other baboons with different behavioral predilections, that this culture can survive. Sapolsky is hopeful, and we cannot fault him for that, but we can note that it does not seem at all the position most supported by the evidence.

Richard Wrangham (2019) offers a perspective on in-group and out-group dynamics that is more consistent with the evidence. He has explored the seeming paradox that humans are, historically speaking, at one and the same time far more peaceful and cooperative inside their own societies and far more warlike and aggressive with other societies than our other primate relatives. Aggression can be understood as consisting of two distinct kinds of behavior, at least in humans. Reactive aggression is in response to a perceived threat or provocation and is precipitated by a significant emotional response. The bar fight between two mildly intoxicated young men that begins with a minor slight and quickly escalates to fisticuffs or worse is reactive aggression. There is no goal here beyond removing the stimulus of the emotional reaction. This kind of aggression is most common in males, and it is often related to male status hierarchy and the relationship that has to a male's ability to attract and keep mates. Much killing in human societies in peacetime consists of examples of reactive aggression.

By contrast, proactive aggression is goal-oriented, planned, and rationally executed. War is the primary example Wrangham gives and dwells on, but many instances of group-based antagonism and conflict with out-groups fit his model well. Wrangham's framework is fully compatible with Alexander's theoretical approach to moral behavior and systems. Reactive aggression—exceptionally high in chimps, our closest living ancestor—was suppressed in human evolution by the emergence of the threat of collectivities of strangers, i.e., other social groups. Chimps also do war—a group-based form of aggression to defend and take territory from invaders and adversaries—but humans have developed this form of in-group/out-group aggression and antipathy to a much higher level. We are, Wrangham writes, both Rousseauian and Hobbesian at once—comparatively peaceful, most of the time in any event, among our brothers, but easy to rouse to suspicion, hostility, and overt conflict with those not in our group.

Racial and ethnic identity would in human history have emerged as fairly straightforward boundary markers between the "us" and the "them." Wrangham, like others who have considered this question, argues that the evidence supports high levels of proactive aggression in human societies even in pre-agricultural groups. Many tribal societies lose as many 0.5% of their populations annually to intersocietal combat, i.e., war, and if this rate were applied to the world population during the twentieth century, it yields a death toll about twenty times higher than the actual rate of death in war for that population during that period (Wrangham 2019:93). Classic examples of horticultural and hunter–gatherer warfare can be found in the literature and ethnographic film of anthropology, e.g., *Dead Birds*, Robert Gardner's celebrated film of the Dani in New Guinea, and the various books and articles by Napoleon Chagnon on the Yanomami at the border between Venezuela and Brazil.

In an American context, the way in which the true history of the wars between European settlers and indigenous peoples of North America belies the mythology of contemporary popular understanding serves as another example. An earlier "good settlers/bad natives" narrative—to be expected, perhaps, based on the rule of ethnocentrism—gave way, at least in some circles, to a full reversal narrative of "bad settlers/good natives." But the inescapable reality is that both groups were fully human, i.e., predisposed to the in-group/

out-group, proactive aggression dynamic and the possibility of escalation from that basis to lethal violence in certain circumstances. Gregory Michno (2003) has extensively documented the depravity on both sides in these wars. The startling image of US cavalryman Sgt. Frederick Wyllyams, killed and mutilated by southern Cheyenne in 1867, is one of the most powerful visual bits of evidence remaining from those bloody conflicts. Wyllyams' lifeless body bears five arrows, numerous other arrow wounds, and it is opened from mid-chest to the groin, with long cuts on each thigh running from the groin to the knee. Indigenous Americans, contrary to the contemporary popular understanding and consonant with the original narrative, frequently scalped war victims, and they did not learn this practice from the settlers. Archaeological evidence shows ancient, pre-Colombian indigenous American skulls with the characteristic cut marks that prove scalping (Keeley 1996:101). The image of Robert McGee, who at age 14 survived a scalping in Kansas by the Sioux, can be found online as evidence of the fact that even the pre-modern indigenous peoples of the Americas, vaunted by many European romantics for their purported peacefulness and aversion to violence, had the same propensities for in-group/out-group aggression that we find in all other humans. Robust evidence of serious warfare between different indigenous American groups before the arrival of Europeans also points to this conclusion (ibid.:109).

Animosities and tensions between racial groups continue to challenge contemporary societies, arguably especially those that have officially embraced some variety of multiculturalism in demographic policy and cultural identity. Nothing said to this point in this chapter indicates that the challenge posed by such racial group conflict is an insuperable problem, but it would be to deny the evidence to say that it does not present in at least some situations as a social issue of great gravity, complexity, and intractability. Knowledge about the similarities and differences between racial groups, and about the causes of disparities in the social outcomes they achieve, is our single greatest resource in trying to address and overcome such challenges. I have already discussed the scientific examination of the causes of racial group differences in IQ in chapter two. The jury can still reasonably be seen as out on the question of the precise genetic nature of group-based differences in IQ, but suggestive data are already coming in on specific genetic loci that seem well correlated with intelligence apart from environmental influences. Numerous other social scientific research areas of some significant social policy implications will likely be affected by ongoing revelations about the biological contribution to human population diversity. How this discussion will evolve will be determined by a number of complex factors, not least of which is the growing phenomenon in some intellectual circles to consider the entire question of the genetic aspects of human group differences as so morally beyond the pale as to make it a taboo subject that should be avoided altogether. Nathan Cofnas (2016) has shown that relying on scientific research on intelligence to objectively sort out the contribution, if any, of genes to human IQ differences will only work if scientists can be counted on to operate scientifically, that is, to research the question and report on results honestly and objectively. He has presented some of the disquieting evidence that substantial numbers of scientists have biases that make this unlikely.

The Mainstream Sociological Analysis of Racial Disparities and the Phantom of "Structural Racism"

An evolutionary theory of racial groups and racial group differences that hypothesizes possible population differences in the distribution of personality and behavioral traits among different clinal groups and empirically investigates such possibility enables a much fuller perspective on the difficult question of accounting for the causes of inequalities and social disparities along racial lines. This latter question is of course a central concern in mainstream sociology, but the approach and the thinking one typically encounters from sociologists here is, to say the least, limited.

It is often presented by professional sociologists and other scholars informed by their thinking on this as a matter of fact that any disparities in social outcomes for different groups can have only one possible cause: Prejudice and discrimination by dominant racial groups against dominated ones. But the causes of social outcomes are almost always complex and multiple. A frequent tactic in mainstream sociology of racial difference—as well as in considerations of disparities in other realms, e.g., gender—is to attribute causal force to vaguely defined "structures" that exceed the level of action of individuals. This move is often motivated by the fact that the evidence for explicit discrimination is paltry or altogether non-existent. By every measure, explicit and damaging episodes of racism and prejudice have been steadily shrinking in the United States over roughly the past half century. All opinion data show a diminution of the purview of hostile beliefs about racial minority groups. At the end of WWII, for example, fewer than half of whites in the US believed in equal opportunity for jobs among the races. By 1972, just a few decades later, nearly all whites (97%) believed it (Krysan and Moberg 2016).

Traditionally, the notion of structure in the social sciences has referred to more or less long-term sets of relationships between different positional placeholders in a society. So e.g., social class is often talked about as an aspect of the structure of contemporary societies insofar as there are typically different strata of positions within the society that are defined by occupational type and/or level of income. This aspect of structure is a long-term feature of such societies. It does not change much over time, although different individuals occupy the different positions at different times, and it can reasonably be thought of as a feature of the society that is substantially independent of the decision-making of individuals. No matter what I or you do in our economic lives, this class structure pre-existed us and will outlive us. But conceptual inventions such as 'structural racism' are another kind of entity entirely. The problem one finds increasingly in American sociology and other cultural spaces influenced by it is that people have started illegitimately expanding the use of the term "structure" to talk about aspects of social positioning in a way that purports to be causal but that simply omits all individual-level causal influence in the production of the positional outcome. Many residential areas, for example, are significantly racially segregated in the United States. What is the cause of a fact like this? It is illegal for communities to explicitly segregate on racial lines, and any housing agencies that were found to be explicitly preventing members of some groups from living in given residential areas solely based on their race would be in legal hot water quickly.

What else is left as possible cause? Income levels of potential homeowners are obviously important. You are effectively prohibited from living in some communities if the typical home there is too expensive for you to afford. Different racial groups have different income distribution patterns, e.g., there is a higher percentage of blacks in the lower income brackets than is true of whites or Asians. This will therefore affect how many blacks we are likely to see in some high-income neighborhoods. But full knowledge of cause is still unavailable at this level. We need to know why there are different income distribution patterns in different racial groups. We know there are things that correlate neatly with income in one direction or another: Education, rates of marriage and intact families, crime rates. On all of these, blacks have a different populational profile than whites. We have only pushed the causal question back another level. Do we fully understand why blacks typically have lower educational attainment or higher crime rates than whites? The short answer in 2020 is: We do not. We certainly cannot simply assert that prejudice is the ultimate cause without sufficient empirical indication and disentangling of other effects. But the "structural racism" argument contents itself with the assertion that once we know there is a differential outcome, we can confidently say that racism is the cause. This is, to put matters simply, not a legitimate scientific conclusion.

A starkly misleading example of this kind of argument in contemporary sociology, widespread and presented to many students in introductory and other courses on sociology, is linked to the functioning of the criminal justice system. In a number of popular works—most centrally Michelle Alexander's *The New Jim Crow*—and in many academic works as well, it is claimed that American society is a seamless system operating according to a logic of "structural racism," perfectly and malignly organized to accomplish the goal of the subordination of blacks. It does this work without the need for conscious intent to make or maintain it on the part of any given individuals, and so no discernibly racist individuals are needed to make the system work. It pursues these goals inexorably and indiscriminately, under its own inevitable momentum and with its own sinister, ghostly mechanisms. Alexander claims that what she is describing is evidence of a vast caste system that dooms blacks to second-class citizenship. A few here or there might escape, by accident or luck, but the purview of the system is universal in theory. The central arm of this caste system is the war on drugs and the effort to imprison vast numbers of non-violent black drug offenders for inordinately long periods of time. Through this malign mechanism, vast numbers of black lives are irreparably destroyed, and all of black America is kept in second-class status as an extension of this mass state violence.

What is the theoretical argument and what are the empirical facts behind the claim? The language of "caste system" implies structural holism and inescapability, an inevitable entrapment of the targeted population. But how many people are directly affected by this massive system? As of 2018, there are roughly some 180,000 people in federal prisons, and about half of these are serving time for drug offenses.[2] There are many more prisoners, just under 1,300,000, in state prisons, but only 14% of these (i.e., around 180,000) have a drug offense as their most serious offense. This gives a total of around 270,000 people in the whole country in prison for drugs—out of a total prison population of just under 1,500,000. This amounts to just under a fifth of the whole prison population.

There were roughly 329 million people living in the US in 2018, so the total number of people in prison for drug offenses was 0.08%, less than one-tenth of 1% of the overall American population. Blacks made up 14.5% of the American population, which in 2018 population figures from the US Census Bureau gives a total of just under 48 million blacks. If we assumed that all the people in prison for drug offenses are black, which is not close to the truth, we would be talking about 270,000 out of 48 million. This works out to 0.6%, or a little more than half of 1%. Alexander expands the number of people affected by her caste system by including not just those currently in prison but anyone currently under correctional control—all those in prison, on probation, and paroled. In 2017, that number stood at around 6.5 million (Kaeble and Cowhig 2018). How many of these are black? Typically, whites and Hispanics together make up more than half the prison population at any given time. They are around 54% of prisoners in the 2018 BJS data cited above, and blacks are about 33%. Roughly the same distribution holds true for the overall population of former prisoners. If we reasonably assume that the portion of those under correctional control who are black is equal to the portion of current prisoners who are black, we get just over two million, which means about 4% of blacks in the caste system, and 96% outside it. But recall that it is the war on drugs and differential sentencing on drug offenses that are the central method of the caste system Alexander describes. Most of those who are under correctional control are not drug offenders and former drug offenders. They are people who have committed other serious offenses—murder, rape, robbery, etc. If we assume the percentage of all those under correctional control who are convicted drug offenders is roughly the same as the percentage currently in prison for drug offenses, we get only a fifth of the total, or 1,300,000 of those under correctional control. If we assume, quite reasonably from current prison data, that only about 33% of this figure is made up of black offenders, that leaves some 430,000 blacks altogether who are in prison, paroled, or on probation for drug offenses. That would be fewer than 1% of all African-Americans in the country under the control of the structurally racist caste system. The overwhelming majority of those under correctional control are male, so if we do this math just for that half of the black demographic, we get just under 2%.

By the very numbers Alexander presents, then, this is the explanatory power of the claimed "structural racism" of what she characterizes as a "New Jim Crow." Only a tiny percentage of the overall black population is directly affected. In scientific terms, of course, the problem is much more significant. There is here no scientific effort to show that "structural racism" is the cause that has produced the encounters these individuals had with the correctional system. That would require careful and comparative consideration of all the factors that distinguish the situation of these individuals from that of others—any evidence of overt racial discrimination against them in the police and judicial treatment of them, of course, but also any factors that might have produced in them a propensity to run afoul of the legal and correctional systems to begin with, e.g., attitudes and beliefs, behavioral predispositions, etc. There is nothing in Alexander's book that weighs any of this.

But she is not a sociologist, a critic might claim. Where is the evidence that this view of racial disparities as inevitably the product of invisible "systemic or structural racism" is propounded by sociologists? I frequently give my students a high profile example in the form of an article by a former president of the ASA, Eduardo Bonilla-Silva. In his "The Structure of Racism in Colorblind, 'Post-Racial' America," he takes a position that is theoretically perfectly consonant with that of Michelle Alexander. The logic of the piece is so lacking in serious investigation of the true causes of racial disparities that my students are often dumbfounded that such work can have been produced from such institutional heights inside the discipline. Faced with undeniable reduction in the level of racial disparities in the United States since the mid-twentieth century, Bonilla-Silva endeavors to explain the remaining such disparities by recourse to a structural form of racism more or less identical to the one Alexander describes. In one example into which I delve in detail in class, Bonilla-Silva discusses a comment by an interview subject who rejects his responsibility for the phenomena of slavery and Jim Crow by claiming a logic of cause and effect that can scarcely be taken seriously by any scientifically-oriented analyst. An assertion of the effect of historical events on present affairs in society requires a demonstration of precisely how the event manifests itself as a causal force in the behaviors of individuals now. Bonilla-Silva claims that evidence takes the form of "the preferential treatment Whites currently receive in jobs, housing, elections, and in almost every area of life" (Bonilla-Silva 2015:1366) and cites another article he wrote with a colleague (Bonilla-Silva and Dietrich 2011) as support for the statement. In neither the article I give students nor the one he cites here does the reader find anything more than a list of disparities without any effort to show what caused them.

What would Bonilla-Silva need to back up the claims he makes? He would have to show not simply that whites have on average better outcomes on some measures than blacks, because those differential outcomes could have to do with myriad factors, but rather evidence of the causes of those outcomes.

Explicit discrimination might be a factor—though, again, it has to be empirically demonstrated, not just claimed as self-evident—but so too might particular behaviors and cultural beliefs that are informing and guiding those behaviors of the groups involved, and they might also have something to do with innate differences between the groups in terms of the distribution of given abilities and skills. Most likely, what is going on is some complicated combination of all or some subset of these. As Bonilla-Silva is forced to admit, outright racist decision-making to keep blacks out of jobs, or out of particular neighborhoods, is now outlawed. What is the evidence of preferential treatment? Typical such efforts in the mainstream of sociology include pointing to studies showing that in at least some situations some people will make decisions about who to interview for a job based at least in part on names and that therefore names that sound ethnically non-white might get treated differently, but that literature is contested. Some have questioned whether the finding is generalizable, and others who accept the reality of the finding point out that the studies show it at a significance that is nowhere close to sufficient alone to explain racial differences in employment outcomes. And an evolutionary framework on such research results might ask what underlies the decisions individuals make about hiring. Stereotyping, about which I will say more below, is almost certainly involved, but its logic and parameters cannot be effectively understood by waving at "structure."

Bonilla-Silva insinuates that structural racism is evident in the fact that middle-class whites typically have negative opinions about mandatory busing that would send their children to schools in poorer neighborhoods. But why is this not simply evidence of individuals acting in a way that is consonant with their own reproductive interests, especially insofar as they might be operating on kin selection in the interests of their children? Crime rates tend to be higher in poorer neighborhoods, and schools in those neighborhoods typically have poorer records of student achievement and discipline. Bonilla-Silva's insinuation is that it is somehow a rationally unjustified decision for a parent to want her child to stay in the school system the parent chose by deciding to buy a home in a particular neighborhood instead of acquiescing to a busing program that would send the child instead to a school with a markedly worse track record regarding student achievement. It is clear here that he is relying not on scientific but on political frameworks of judgment. The insinuation he makes is that not being a color-blind racist would require parents to acquiesce to whatever political platforms Bonilla-Silva has decided are the optimal platforms for making society better. If one agrees with his politics, one might find that a convincing political position, but it is certainly not a convincing way to build a reliable social scientific theory about racial groups, racial identity, and human behavior.

Bonilla-Silva's argument can be neatly summarized: Slavery existed, and then Jim Crow, and this means that white supremacy never really died, no matter how the situation with respect to racial hierarchy has demonstrably and radically changed since the 1960s and even though racism had to transform itself, and so "color-blind racism" —which coincides with the ghostly continuing presence of slavery and Jim Crow—is the sole and exhaustive cause of all racial inequality. But the reasons why people are employed or unemployed, offered mortgages or denied them, admitted to college or not, imprisoned or acquitted, are numerous and complicated. If explicit prejudice has demonstrably receded, what is left in the way of empirical indicators? What of the possible role of racial differences in educational training, testing performance, and evidence of income and savings, wealth, and accumulated debt, or of anecdotal or perhaps more systematic employer knowledge of differential job performance between groups, to name just a few other potentially relevant variables? Why do black students drop out of school more frequently? Why are their test scores lower? Why do they have less success securing good jobs? Why are they overrepresented in the criminal justice system as defendants? The very business of truly scientific social analysis, of exacting and difficult examination of social problems with objective methods and theoretical equanimity, is side-stepped by analyses like that of Bonilla-Silva. The promise of an evolutionary framework for understanding group identities and differences is not that it will reduce causes to one level, but that it puts back in a set of potential factors that were inexplicably and unscientifically excluded as a matter of ideology by mainstream sociologists.

A better effort to at least begin to theorize how historical events and institutions that no longer exist can nonetheless affect what happens in the present can be found in the work of Orlando Patterson (1999). Patterson argues that, during slavery, there was a systematic effort to keep black men from fully inhabiting and exercising key aspects of the roles of husband and father. Although black men and women were allowed to partner and to have children, a strenuous effort was made to constantly remind black men that they could not prevent other men from having sexual access to their wives and they were helpless to stay with their children if their owners decided to break up families. Patterson argues that black men living in these conditions adjusted their behavior and cultural values with respect to sex and children by downplaying the roles of husband and father and instead placing value on a male persona that was sexually potent but unconcerned about emotional investment in women or children. The problem is that though this cultural set of values and practices was rational and probably psychologically protective during that historical period, and though it has been moderated by the passage of time, it is still a value system that informs the behavior and attitudes of many, and it is now destructive to black community and achievement. This shows the enduring effects of slavery in the empirical statistical patterns in mating and marrying among blacks, and in this sense the consequences of the institution of slavery are still with us. However, Patterson is theoretically careful and sophisticated enough to recognize that this means that we are talking about the effects of history empirically as they are present in cultural systems and behavioral patterns among blacks that are informed by those cultural systems. It is too simple to say that this is being imposed in the present day by "color-blind" or "structural" or "systemic racism" or some other such vague concept. These cultural systems of meaning surrounding marriage, sex, parenting, etc. are in Patterson's analysis effective precisely because many blacks actively embrace them as aspects of their meaningful worldviews, and there is therefore no need for whites to impose them today in order for them to sustain themselves. Patterson does not argue from an evolutionary framework, and so the ultimate level of analysis—which underlies even human culture—escapes him, but his work is certainly far closer to the kind of scientific sociology of racial disparities we need than we find in more mainstream work like that of Bonilla-Silva.

Generalizations, Stereotyping, Human Group Categorization

Humans are inveterate classifiers. We place discrete things into groups as a matter of deep cognitive predilection and necessity. We are, with appropriate levels of knowledge, perfectly capable of perceiving and classifying by recourse to individual difference and uniqueness, but we are also adept at classifying by observed and inferred similarities of groups of individuals. Stereotyping has in individualistic societies like those in the modern West taken on a pejorative tint, and it is assumed by many educated people that it is an entirely illegitimate and unjustified way of making sense of the world, in all cases. Yet it is undeniable that it is a potentially useful skill, and its use in our species can be effectively understood from the natural selection logic employed in this book. Some stereotypes are of course inaccurate, and some of these can provide false legitimations for behaviors that are harmful to the individuals who are stereotyped. But this is not so of all stereotypes. Some are based in a reasonably accurate comprehension of the characteristics of individuals, groups, and situations. When you think a minute about this in an evolutionary perspective, it is not hard to see why this is not unlikely. How would a general propensity to this kind of behavior have emerged and persisted if all stereotypes were wholly false and therefore acting on them would have likely produced no benefit for those who did so?

Stereotypes are generalizations, so we should not expect them to be unfailingly accurate. For example, it is a stereotype that men are more physically aggressive than women, who are stereotyped as more compassionate than men. Many exceptions to both stereotypes—passive, unaggressive men and uncompassionate, combative women—obviously exist. Voluminous evidence, however, demonstrates the basic congruity with reality of the stereotype as a statement about statistical means of the two populations. On average, men are more aggressive and women are more compassionate, and if you knew nothing about a random man and a random woman other than their sex and were asked to pick which one was more aggressive and which was more compassionate, the safe bet would be in line with the stereotypes. Even many stereotypes elite culture today classes as offensive are based in empirically accurate generalizations. The belief that blacks and Latinos commit violent crime at higher rates than whites and Asians is true, though, again it does not follow that all members of the first two categories and none of the last two are criminals. The theory of natural selection might suggest an interpretation of the phenomenon that runs roughly thus: Early hominins, and later modern humans, who stereotyped in their interactions with others might have gotten things wrong a considerable amount of the time, yet the stakes might well have been high enough in some situations—e.g., stereotyping a stick shape in the bush as a snake, or a stranger as a dangerous enemy rather than a potential friend—to provide advantage to those who used the technique, and thereby to invoke the mechanism for reproducing it in descendants, perhaps both in terms of heredity—that is, by selecting for any genetic variant contributing to a personality type and behavioral predilection in this direction—and cultural socialization.

In modern societies, we often meet individuals about whom we know almost nothing in personal terms. What knowledge are we likely to use in understanding them and our likely interactions with them? Part of the stereotypical tool kit, if you will, in modernity involves a baseline, unexamined trust that people one does not know are much like oneself—mostly peaceful, and in any event not interested enough in adjacent strangers to bother them. But there are other parts of this toolkit as well, some of which have to do with in-group/out-group perceptions and information that are probably rooted quite deeply in our natures. It is not hard to find convincing rationales for using them and good data that back up their use in at least some situations. Imagine, for example, that one is a parent interested in hiring someone to provide childcare at one's home. Two candidates present for an interview. One has multiple tattoos and body piercings. If one had deep personal acquaintance with both people, of course, that bit of data might not play into the decision in any substantive way at all. But imagine one does not know either candidate beyond what is on their applications and what one sees, which is the real-world situation typical of most such hiring scenarios. Many hold the stereotype that people with body piercings and tattoos merit more relational caution and suspicion than people who lack such bodily modifications. But is there any real basis for this stereotype? It turns out that there is a literature demonstrating a positive relationship between tattoos and body piercing, especially multiple tattoos and/or piercings, and a range of negative personality traits—low constraint, i.e., impulsive thrill seeking, and callous affect, expressing anger verbally, high negative and low positive emotionality generally—and behavioral predilections—alcohol use, smoking, drug use, high-risk sexual behavior, anti-social activity (see e.g., Bui et al. 2010; Jennings, Fox and Harrington 2014; Ruffle and Wilson 2019).

Given this information, are we prepared to assert that it is wholly unreasonable in a situation such as the one described to act on the stereotypical information about people with body piercings, given the stakes, that is, that the individual in question will be in unsupervised contact with one's minor child? Whether we personally accept the reasonableness of using such a stereotype here, could we reasonably expect others, as a matter of moral insistence, not to avail themselves of such information in making such decisions, again, absent other and personal information that is often denied to individuals in their day-to-day lives in mass societies? Surely there are at least some scenarios involving, for example, decision-making about what cities, towns, and neighborhoods to live in and what schools to which to send one's children that will involve stereotyping with at least some grounding in empirically true generalizations about different groups. Given that stakes are high in some such scenarios, stereotyping about groups is thus unlikely to disappear altogether, even with stern top-down efforts to detect and punish such decision-making models. We certainly have a collective interest in preventing illegitimate and empirically unsound such stereotypes from producing clear harms for individuals, and nothing argued here implies abandoning all efforts to prevent such outcomes. Moral and political programs as to what to do about practical and behavioral manifestations of in-group/out-group are to some degree independent of science as a way of knowing, and such different frameworks for morally considering the complex questions involved will foresee different political solutions. But a proper science of group differences would at least provide all of us, whatever our moral and political commitments, with the most empirically rigorous understanding of the causes of particular acts of homologous affiliation and heterologous contraposition and thereby make us better informed as to the effects of such behaviors and the most feasible ways to act upon them through social policy.

This example illustrates the complexity and the challenge of reasoning and thinking about human difference. We know humans make distinctions based on phenotypical differences, and we know they sometimes act on them in a wide range of ways. Surely knowing the details of how and why our species thinks about and acts on such differences would be useful knowledge. Modern ethnically and racially pluralist societies have a great interest in finding ways to make it possible for diverse members of those societies to live together in the greatest harmony possible, but the empirical evidence is that this is a project with great challenges, to say the least. We might well wish that were not so, and that it were possible to solve such difficulties through the invocation of simple moral principles and hand-waving at scientifically-untenable explanations for conflicts and disparities. But what if these prove insufficient as solutions to the difficulties such societies face? There is no doubt that there are real questions and concerns to be raised, and we must proceed cautiously and judiciously. But it is equally certain that the science of fully understanding the basis and nature of human populational differences at all levels—social, cultural, and genetic—is an essential element in the construction of the kinds of social systems that endeavor to reduce the most serious kinds of conflict among diverse groups with a realistic view of basic features of human nature.

At the scientific and professional levels, sociologists face a choice: Support such scientific inquiry and do the homework required to get directly involved in it, and thereby have some influence in how it is discussed in the public sphere and put to use in social policy, or reject and denigrate it from a position of demonstrable ignorance, and watch from the sidelines as other, more rigorous, more serious disciplines produce the research that influences policy and the evolution of society.

Notes

1 http://www.orderofthegooddeath.com/how-to-fix-society-one-baboon-group-at-a-time.
2 All data in this paragraph are from the 2018 US Department of Justice report on prison statistics (Carson 2020).

References

1000 Genomes Project Consortium, The. "A Global Reference for Human Genetic Variation." *Nature*, 526, 7571 (Oct. 1, 2015):68–74.

Alexander, Michelle. *The New Jim Crow: Mass Incarceration in the Age of Colorblindness*. New York: The New Press, 2010.

Bonilla-Silva, Eduardo. "The Structure of Racism in Color-Blind, 'Post-Racial' America." *American Behavioral Scientist*, 59, 11 (2015):1358–1376.

Bonilla-Silva, Eduardo and D. Dietrich. "The Sweet Enchantment of Color-blind Racism in Obamerica." *Annals of the American Academy of Political and Social Science*, 634 (2011):190–206.

Bui, Eric, Rachel Rodgers, Lionel Cailhol, Philippe Birmes, Henri Chabrol, and Laurent Schmitt. "Body Piercing and Psychopathology: A Review of the Literature." *Psychotherapy and Psychosomatics*, 79 (2010):125–129.

Carson, E. Ann. *US Department of Justice: Prisoners in 2018*. April 2020 https://www.bjs.gov/content/pub/pdf/p18.pdf.

Cavalli-Sforza, Luigi Luca. *Genes, Peoples, and Languages*. Translated by Mark Sielestad. Berkeley: University of California Press, 2000.

Cavalli-Sforza, L. Luca, Paolo Menozzi, and Alberto Piazza. *The History and Geography of Human Genes* (abridged paperback edn). Princeton: Princeton University Press, 1994.

Cofnas, Nathan. "Science Is Not Always 'Self-Correcting': Fact–Value Conflation and the Study of Intelligence." *Foundations of Science*, 21 (2016):477–492.

Conley, Dalton and Jason Fletcher. *The Genome Factor: What the Social Genomics Revolution Reveals about Ourselves, our History, and the Future*. Princeton: Princeton University Press, 2017.

Coyne, Jerry. *Why Evolution is True*. New York: Viking Press, 2009.

Dobzhansky, Theodosius. *Mankind Evolving: The Evolution of the Human Species*. New Haven: Yale University Press, 1962.

—. *Genetic Diversity and Human Equality: The Facts and Fallacies in the Explosive Genetics and Education Controversy*. New York: Basic Books, 1973.

Dobzhansky, Theodosius, F.J. Ayala, G.L. Stebbins, and J.W. Valentine. *Evolution*. Freeman: San Francisco and Reading, 1977.

Edwards, A.W.F. "Human Genetic Diversity: Lewontin's Fallacy." *BioEssays*, 25 (2003):798–801.

Fujimura, Joan H., Deborah A. Bolnick, Ramya Rajagopalan, Jay S. Kaufman, Richard C. Lewontin, Troy Duster, Pilar Ossorio, and Jonathan Marks. "Clines without Classes: How to Make Sense of Human Variation." *Sociological Theory*, 32, 3 (2014):208–227.

Garn, S.M. *Human Races*. Springfield, IL: Charles C. Thomas, 1961.

Golash-Boza, Tanya Maria. *Race and Racisms: A Critical Approach*. New York: Oxford University Press, 2015.

Gould, Stephen Jay. *The Mismeasure of Man*. New York: W.W. Norton & Company, 1981.

Hartshorn, Max, Artem Kaznatcheev, and Thomas Shultz. "The Evolutionary Dominance of Ethnocentric Cooperation." *Journal of Artificial Societies and Social Simulation*, 16, 3, 7 (2013).

Jennings, Wesley G., Bryanna Hahn Fox, and David P. Farrington. "Inked into Crime? An Examination of the Causal Relationship between Tattoos and Life-Course Offending Among Males from the Cambridge Study in Delinquent Development." *Journal of Criminal Justice*, 42 (2014):77–84.

Kaeble, Danielle, and Mary Cowhig. *US Department of Justice: Correctional Populations in the United States, 2016*. April 2018 https://www.bjs.gov/content/pub/pdf/cpus16.pdf

Keeley, Lawrence. *War Before Civilization*. New York: Oxford University Press, 1996.

Krimsky, Sheldon and Kathleen Sloan. *Race and the Genetic Revolution: Science, Myth, and Culture*. New York: Columbia University Press, 2011.

Krysan, M., and S. Moberg. *Trends in Racial Attitudes*. University of Illinois Institute of Government and Public Affairs. August 25, 2016. http://igpa.uillinois.edu/programs/racial-attitudes

Lee, Sang-Hee, with Shin-Young Yoon. *Close Encounters with Humankind: A Paleoanthropologist Investigates Our Evolving Species*. New York: W.W. Norton & Company, 2015.

Lewis, J.E., D. DeGusta, M.R. Meyer, J. M. Monge, A.E. Mann et al. "The Mismeasure of Science: Stephen Jay Gould versus Samuel George Morton on Skulls and Bias." *PLoS Biol*, 9, 6 (2011).

Lewontin, Richard. "The Apportionment of Human Diversity." *Evolutionary Biology*, (1972): 381–398.

Lieberman, Leonard, Katarzyna A. Kaszycka, Antonio J. Martinez Fuentes, Leonid Yablonsky, Rodney C. Kirk, et al. "The Race Concept in Six Regions: Variation without Consensus." *Collegium Antropologicum*, 28, 2 (2004):907–921.

Lopreato, Joseph. *Human Nature and Biocultural Evolution*. Boston: Allen & Unwin, 1984.

McEvoy, Chad Joseph. "A Consideration of Human Xenophobia and Ethnocentrism from a Sociobiological Perspective." *Human Rights Review*, (April–June 2002).

Michno, Gregory. *Encyclopedia of Indian Wars: Western Battles and Skirmishes, 1850–1890*. Sevierville, TN: Mountain Press, 2003.

Omi, Michael and Howard Winant. *Racial Formation in the United States*. New York: Routledge, 1986.

Patterson, Orlando. *Rituals of Blood: Consequences of Slavery in Two American Centuries*. Washington, DC: Civitas Books, 1999.

Pinker, Steven. *The Blank Slate: The Modern Denial of Human Nature*. New York: Penguin Books, 2002.

Reich, David. *Who We Are and How We Got Here: Ancient DNA and the New Science of the Human Past*. New York: Pantheon, 2018.

Ruffle, Bradley J. and Anne E. Wilson. "Tat will tell: Tattoos and Time Preferences." *Journal of Economic Behavior & Organization*, 166 (Oct. 2019):566–585.

Sackett, P.R., C.M. Hardison, and M.J. Cullen. "On Interpreting Stereotype Threat as Accounting for African American–White Differences on Cognitive Tests." *American Psychologist*, 59 (2004):7–13.

Sapolsky, Robert. *Behave: The Biology of Humans at our Best and Worst*. New York: Penguin, 2017.

Shiao, Jiannbin Lee, Thomas Bode, Amber Beyer, and Daniel Selvig. "The Genomic Challenge to the Social Construction of Race." *Sociological Theory*, 30, 2 (2012):67–88.

Tang, Hua, Tom Quertermous, Beatriz Rodriguez, Sharon L.R. Kardia, Xiaofeng Zhu, et al. "Genetic Structure, Self-Identified Race/Ethnicity, and Confounding in Case-Control Association Studies." *American Journal of Human Genetics*, 76, (2005):268–275.

Tishkoff, Sarah et al. "The Genetic Structure and History of Africans and African Americans." *Science*, 324, 5930 (2009):1035–1044.

Wrangham, Richard. *The Goodness Paradox: The Strange Relationship between Virtue and Violence in Human Evolution*. New York: Pantheon, 2019.

8 Culture in *Homo sapiens*

The topic of this final chapter is the one I have written most on in my professional life as a sociologist to date, yet I treat it more briefly than several others discussed in the book. The point I hope to make with such comparative brevity is this: Culture is much less important in understanding why human beings do what they do and how they organize their societies than we might like to think. I deliberately use the collective subject form because I long was one of those who believed it central to that task. I wrote a fair amount in the cultural sociological vein over the course of the first fifteen years of my scholarly career. Two of my previous books are neatly situated in this subfield of sociology. In the first of these books, the topic was transgression (Riley 2010); in the second, the memorialization of catastrophic events (Riley 2015). In both, I endeavored to describe as much of the complicated cultural narrative surrounding these two forms of behavior as possible, including conflict over the meanings of such behaviors.

But how much in either of those two studies can reasonably be seen as explanation of those behaviors? Neither attempts to explore how the theory of natural selection might underlie the proximate causes examined. Transgression, I argued, is functionally motivated by the very normative system against which it transgresses. Why does a social system require such behavior? Because of the psychology of the people in it. But what in our psychology motivates transgression? Natural selection would suggest that if there is any significant amount of it—and the evidence suggests this is so—it must offer some selective advantage. Sexual transgressions of restrictive normative systems seem an obvious case to examine. Does sexual promiscuity enhance reproductive success, especially for males? In the social environments in which we in the West currently live, wrapped as we are in moral codes and coinciding legal systems that stigmatize and punish transgressors, likely not. But in other ecologies, and especially in the kinds of settings in which humans have spent most of our time on the planet, in which moral rules and institutions to restrict and punish it were weak or non-existent? Perhaps so.

In the case of memorialization, the evolutionary social scientist would perhaps want to trace such cults of memory back to their beginnings in the original such cults, that is, the earliest human practices of burial of the dead. Clearly, religious motivations might be provided at some point in the emergence of this process to explain it—we bury the dead because the gods require us to do so. But what about ultimate causation? Why do we have gods that require us to commemorate our dead? I referred frequently to Durkheim's interpretation of religion as a symbolic means by which societies in effect worship themselves, recognizing at least implicitly the superior power that is the collective over the life of the individual. I did not go further, though, to speak to the base-line question of why human collectivities would produce such masked forms of veneration of the group itself. Richard Alexander's discussion of morality, which I presented in chapter four, points toward an answer: Moral systems that involve reciprocal altruism are effectively bolstered by such forms of belief.

These would have been two rather different books if I had written them while wearing the lenses I wear as I write this one. I do not mean to imply that I believe those two books are wholly without any utility, as I should certainly like to hope it was not a complete waste of my time to write them! But they do not even potentially offer explanations of the behaviors described. They describe and they meticulously work through the stories human groups tell themselves about the actions described, but they do not explain. If explanation is to be our business as social scientists, as I have argued in this book it must be, then we must do more.

Culture Exceeds the Human

Only humans "do" culture. It's what makes us human, what distinguishes us, what elevates us. This is an assertion made in the typical textbook that introduces first-year college students to the social sciences. It is challenged by a significant body of evidence from studies of non-human animals. Chimpanzees have been shown to use tools for a wide range of tasks, including procuring difficult to access food such as termites in mounds and cracking nuts, and specific aspects of their tool use vary by population, i.e., there are local cultural groups that differ in the specifics of their tool practice from other such groups. As I discussed in chapter seven, Robert Sapolsky presents evidence of an important population specific difference in grooming in a troop of baboons he observed. Much of the body of data on non-human culture comes from primates, but a growing field of such work involves culture in cetaceans. These animals are harder to study, as they live in oceans that make it more difficult for human scientists to follow them around, but we now have a good deal of evidence that suggests some cetacean species might well be even more cultural than chimps or any other primates. One population of bottle-nosed dolphins in a bay in Australia was found to use sponges secured to the end of their snouts to scour the ocean floor more effectively for a particularly nutritious variety of bottom-dwelling fish. The sponges apparently protect their snouts from injury from rocks and coral on the sea floor. Intriguingly, the cultural practice is largely restricted to female dolphins, who teach the technique to their female offspring but not to males.

A remarkable book written by two marine biologists (Whitehead and Rendell 2015) details another fascinating example. Populations of humpback whales have been studied in their migration patterns between Antarctica and the Coral Sea. The route requires them to go around the continent of Australia, and two large social populations are divided by the direction they take around the huge obstacle. It has been discovered that the song the whales produce differs between the two groups at a structural level, and individual whales have been witnessed to switch songs depending on which migration route they take. They learn the western version when they go that way, and the eastern version when they go that way, rather like the American whose job takes her to Great Britain and who over time develops something of an imitated English accent. The cultural meaning of whale song is not yet entirely understood. Only male whales sing, and the existing evidence suggests that it is related to mating. Singing occurs during migrations and while gathered in the winter breeding areas, but not during the summer while the whales are feeding. There is no clear evidence that females are particularly attracted to singing males, but it does appear that singing makes females receptive to mating generally. Groups of males often sing together prior to setting out collectively in search of females with whom to mate, so it appears that the function of the song is communicative in a dual sense: An announcement to females that helps encourage their receptivity to mating, and an element in the mutual cooperation of males as they prepare to engage in mate pursuit.

I hear the groans of my cultural sociologist and anthropologist friends. Surely here is the clue to why this cannot at all be compared to human musical culture, since it reduces to sex, while human music is so much richer and more complex than mere games of sexual attraction. I certainly grant that there is human music about virtually every conceivable topic. But have you ever wondered how much of it is about sex and related topics? There are arguments about the origin of music—and art more generally—that connect it precisely to sexual selection tactics in a wide range of animals that involve practices such as vocal elaborations, dances, decorative activities like that of the male bower bird, and the other aesthetically compelling display that members of the opposite sex find attractive (Dutton 2010; Boyd 2010). And certainly a great deal of human music is explicitly about sex, from the suggestive content of lyrics to the pulsating rhythm of most pop music. We perhaps do well then not to go too far in idealizing the content of human culture to separate it from the "vulgar" function of whale song.

What is Culture and What is it for?

Whether or not we can put whale song and human popular music in the same category depends largely on the definition of the category. How should we best define "culture"? The definition that sociologists and cultural anthropologists often give to the term builds human uniqueness into it before the fact. Cultural anthropologists are particularly speciesist in this regard, building exclusivity into their entire field. Edward Tylor's definition is one of those most widely disseminated by introductory textbooks in sociology and anthropology, and it gives the humanist game away: "that complex whole which includes knowledge, belief, art, morals, law, custom, and any other capabilities and habits acquired by man as a member of society." A more scientifically useful—and certainly more empirically open-ended and non-anthropomorphic—definition comes to us from the study of whales and dolphins mentioned above. There, culture is defined as "information or behavior—shared within a community—which is acquired from conspecifics through some form of social learning" (Whitehead and Rendell 2015:12). Here we have two marine biologists who have thought more clearly and in a more interdisciplinary, holistic way about culture than most social scientists. We might do well, though, to push the definition still more specifically in one final tweak, following Alex Mesoudi (2011) by defining culture solely as information with those characteristics. If it is behavior too, as Mesoudi notes, then culture and behavior are not distinct, and if we are trying to explain behavior, culture cannot conceivably do the job, since it is the same thing.

What is cultural information for? This is another question seldom asked in mainstream sociology, or one too typically assumed to have a self-evident answer. Culture is for meaning, we are told, because humans require it. Perhaps we do. Unasked here, though, is *why* we do. Let me make a brief attempt at an answer. Culture increases the likelihood that we will get some behavior from people that we class as "moral," that is, not completely self-interested and wholly disdainful of considerations of the needs and interests of others and of the entire social collectivity. Getting humans to cooperate, especially in ways in which it might not be easy for them individually to see how they benefit from such cooperation personally, is not the easiest thing to accomplish, and human societies need to achieve this goal if they are going to cohere and remain competitively competent among other societies. Culture often compels such cooperation through the symbolic means of gods and myths, which as Durkheim noted stand in for the social group itself. The profound energy of contact with sacredness is a hugely useful resource for directing men and women to acts of sacrifice and heroism that they might otherwise eschew utterly. Culture is a bag of tools, in the view of the sociologist Ann Swidler, that are useful for successfully solving problems, especially problems that are particularly troublesome to everyday life and that tend to recur. Culture here is a kind of receptacle into which we put the accumulated body of solutions to such problems, so they will be readily available to members of the social group when they encounter those problems. Reinventing the wheel is a tremendously difficult and time-consuming task in many situations, and culture enables us to save the time and energy we would need to reinvent wheels and instead potentially spend those resources on novel problems. Culture is also a collection of games that serve as outlets for basic human impulses to expend energy and compete with one another, and such cultural games—political election campaigns were on my mind as I wrote this in an American swing state in October 2020—serve as broadly legitimately viewed means of determining status rankings in a hierarchy.

On the Complexity and Location of Human Culture

It is only a kind of extremely blind self-love that makes us believe that human beings are the unbelievably complex and competent cultural machines that appear to us in cultural anthropology and cultural sociology. The mainstream model would have it that culture is both a complicated system of narratives and symbols, and that it is situated fully inside the heads of individual members of the culture. The sociologist John Levi Martin argues compellingly that these cannot both be true

> for the simple reason that our minds are not good at holding lots of connected things in them. If one wants to define culture as something complex, then it is not going to be inside of people…because people are extremely simple.
>
> (2010:229)

This is exactly right. Disappointing as it might be for us to accept, the evidence is strong that people are not nearly as complicated as the interpretive social sciences would have it, at least not insofar as social scientific study of their behavior is concerned. It may be that some considerable portion of any given literate society could, if asked, provide to a questioner at least some of the cultural symbols and narratives that are catalogued by culturalist social scientists. I confess I am less than confident even about that. But I am quite convinced that almost none of the members of any human society is actively using that information to make decisions about more than a small percentage of their behavior. Cultures can certainly consist of highly dense and complex bodies of information, but almost no real person is at all interested in the kind of Geertzian cultural analysis that makes up anthropological ethnographies of culture, nor do many of us have much at all of any of that kind of information inside our heads. Indeed, even most intellectuals are not so complicated either. Any honest member of a university faculty will admit that when observing discussions and debates among groups of college professors, either in person at faculty meetings or in online communication, it is perfectly possible with sufficient experience with that group to predict the basic framework, and sometimes even the content of the interventions of given colleagues on a given topic. I do this constantly, and fairly reliably, with the faculty listserv at my institution. If the topic is A, it is almost inevitable that Professor B will at some point say C or something close to it, as this is what he always does, and then Professor B's nemesis Professor D will respond with some variation on E, which is essentially the response to this stimulus, and this will prompt Professor F, Professor B's ally, to intervene with G, and Professor D's ally, H, will then intervene in a broadly predictable way, and so on. The content can of course be varied, and it at least sometimes—but by no means always! —gives evidence of complexity, given that the interlocutors are college professors who are practiced verbal artists, but the content is arguably not the most interesting sociological piece of the behavioral dynamic. That has to do with the basic game of friend–foe and the network of alliances and animosities, a very basic element of primate behavior. I do not intend any insult of the participants in such interaction dynamics in making this observation, but the point is close to undeniable if one has any reasonable body of empirical data on this.

Perhaps inside their heads, late at night, on weekends, people have a lot of things going on. I have the impression this is true in my own experience, so it is a reasonable guess that it is generalizable. But the action taking place between people is for the most part not nearly so complex. A lot of that is reasonably reducible to the basic games of pursuit of reproductive opportunities, the seeking of positions in a status hierarchy (to pursue better reproductive opportunities and to ensure one's offspring do well), and the acquisition or protection of other resources of some utility in the game of increasing reproductive opportunities. None of this means that cultures are not complex, nor does it deny that this cultural material is fun for us to discuss, at least occasionally. Intellectuals like to write articles and books about it and share those with others interested in such things. But how much does such discussion help us to understand why people do what they do, or how they organize themselves socially? At best, these are discussions of proximate causes. A moral value related to a religious principle enjoining members of the cult to give alms, communicated to a human subject over the course of his socialization as a member of the group, is presented as the explanation for his giving alms. But what produced that moral value? The religious system of the cult, it will be said. And what produced that?

If culture were in our heads, this would be verifiable empirically by simply asking individuals to recount given cultural narratives or systems of meaning from their social groups. You can do this research on your own. Pick a cultural narrative as provided by your favorite sociologist, or historian, or literary critic, and then try to find evidence of it in the articulations of existing human subjects. I routinely ask my students about such narratives. How many can tell me about the stories that cultural sociologists say are driving and giving meaning to their lives? What, I ask them, is the moral basis of American culture? How does American individualism work as a cultural system? Almost all of them are relatively inarticulate when asked such questions. Many are completely flummoxed. Those who do have a response typically crib from some book they have recently read, perhaps in my or another class. How long will such information stay in their heads? Probably not long. They do not need such information to function as members of the group because most of the important material they need for that is much simpler and more basic, and immediately available without conscious effort.

Culture, then, is not in heads. It is elsewhere. But where? It is largely to be found in objects, in the form of material culture, where it can conceivably be tapped into by members of a cultural group, though most of it is unused altogether. I write this sitting in a room filled with books, many of which I have read at least in part. How much of that information is retrievable from my consciousness, without the material aid of the book and the notes I have written in its margins? A tiny percentage. Think of all the books that are sitting in all the libraries, unread by anyone except the author and perhaps the editor, both of whom are dead in many cases. Where is the cultural information in those books to be found in real human interactive life as it plays out before us daily? Unless it has already been decoded and put into some living human behavioral process—a ritual, perhaps—it is nowhere else but in the material form of those books, and unrealized and meaningless until a human brain reads them and remembers them. A good deal of culture is probably to be found in sets of what Pareto would have called derivations, bits of stories and snatches of meaning with which we try to stitch the pieces of our behavior in everyday life into a coherent garment. Below the derivations, or invocations of culture, are the real motivations that drive most of what we do, frequently and perhaps even most of the time, in acts of self-deception so marvelously contrived and formulated by forces beyond us that we do not have the slightest idea that this is what we are doing. There is evidence that there are cultural pre-sets or perhaps what evolutionary psychologists call "modules" for prioritizing cultural knowledge of a kind that is of greater likely utility for selection in deep evolutionary time. For example, children in different environments, modern Los Angeles (where this would seem of no utility) and in the Ecuadorian Amazon (where it might well be of greater use) learn with equal facility about dangerous animals (Barrett and Broesch 2012).

Human Cultural Diversity

In addition to claiming that culture is uniquely human, culturalists frequently argue that all human cultures are unique, that culture is nearly infinitely diverse, and therefore that it is impossible to reasonably talk about cultural types or stages. At the same time, the human species is typically understood as one and unified and there are no meaningful differences among human groups that are not reducible to culture and patterns of socialization and norms. These claims run headlong into the fact of human cultural universals, which are voluminous in number. The anthropologist Donald Brown (1991) compiled a working list of ten pages of such universal features including, to pick just a few, the admiration of generosity, male domination of the political realm, and a preference for one's own children and close kin. All societies, for example, have food and speech taboos, and in every human society about which we have any knowledge, existing or historical, men are more aggressive and violent on average, as I discussed at length in chapter five. In the midst of a long discussion of this topic, Lee Cronk (1999) gives a more mundane example that is telling precisely because of the way it illustrates how much of the possible variation of human culture is not actuated in any existing culture. There are no known examples of amitalocality—the generalized practice of a married couple going to live with the wife's father's sister's family—but its opposite, avunculocality—married couples living with the husband's mother's brother's family—is common. Cronk goes on to provide a compelling mathematical illustration. Using cultural variables given in the Ethnographic Atlas, we find that the number of possible combinations of all these variables is 12 x 10 to the 51st power. The overwhelmingly vast majority of these possible combinations are not realized in any societies (ibid.:23–24).

Cultural anthropology is often misleading, at best, in what it has to say about human nature and behavior. It overemphasizes cultural variation and the uniqueness of given societies and is often fundamentally misleading, even mendacious, on details of cultures. Claims of nearly infinite cultural plasticity are frequently made by culturalists using false examples. The Hopi, it is argued, have no concept of time, so time cannot be a cultural universal. But this is not so (ibid.:21). Symons mentions a particularly egregious case, Ralph Linton's "ethnography" of the Marquesas Island people, which was shown by subsequent research to be a report on "a culture that never existed" (1979:70). It can be argued that this bias, this desire to see uniqueness and to overlook similarity, is an essential element of the culture of professional cultural anthropologists, something into which they are trained from early on in their institutional formation as anthropologists.

Origin of Human Culture

Curiously, much cultural social science evinces little interest in a systematic understanding of the origins of human culture. I looked through a few best-selling textbooks in cultural anthropology and sociology without finding any real clues on this important point. I did encounter plenty of repeated clichés—culture is to humans like water is to a fish—and reiterations of the same broad claims—culture is a complex system, hard to define but evidently essential to being human—but little discussion of where culture could have come from and why it came to be in the first place.

Antonio Damasio gives an account of the origin of culture that is deeply consonant with the basic evolutionary characteristics of life on the planet. What describes life, most fundamentally? It is the "precise mandate [to] resist and project itself into the future, no matter what" (2018:31), or, in the words of the French poet Paul Éluard, "le dur désir de durer…the 'determined desire to endure'" (ibid.:36). All life seeks to extend itself outward in time from the present, and the structure of life on earth provides the mechanism by which that is accomplished. Of course, genetic replication is an essential part of this story, but Damasio suggests the theory that replication is the basic force driving the earliest life is less convincing than an alternative explanation that puts homeostasis first. The physicist Erwin Schrödinger, writing before the discovery of the structure of DNA, defined life as "existing order that is kept up," which Damasio suggests indicates that homeostasis is the primal imperative of life. Early life, as all subsequent life, would have needed methods for identifying the emergence of disordered internal states in order that they could be corrected.

Early single-celled life, however, certainly had nothing like what we call culture. Or did it? Even bacteria communicate with one another and coordinate behavior at least at the basic level of pursuing independent existence or gathering with others of their kind. Neurons and eventually nervous systems emerge much later to carry on this work that is already being done at the inception of life. Nervous systems operate fundamentally as generators of emotional reaction to the metabolic state of the organism. But all the machinery to drive such innate, homeostasis-preserving behavior preexists nervous systems, brains, and emotions. The argument is that culture emerged from animals concerned fundamentally with homeostasis, and so we should seek its rootedness in the drive to maintain that quality. Given that the origins of culture vastly precede in time the emergence of consciousness, cognitive tools and rational decision-making strategies cannot be the driver. Damasio finds—and here he is perfectly in tune with the analysis of Turner and Maryanski that I discussed in chapter three—that culture has its roots in the emotional lives of humans, which depend on neurological and other biological systems they share to a significant degree with other mammals. Emotions are ultimately motivated as responses to organismal threats and disequilibria, and cultures are means for channeling emotional response collectively in directions that can effectively address the failure of homeostasis.

Human Culture and Natural Selection

It is commonly understood by social scientists and some others that human culture has allowed us to escape the causal machinery of genes and natural selection. The part of culture that is science has certainly given us knowledge and tools that allow us, for the first time in the history of the planet, to directly manipulate the genome in ways far more precise than any of the practical methods we have previously used. But this is not what culturally-focused social scientists mean when they speak of humans as outside natural selection. They mean to refer to the ability to know about reproductive fitness and to deliberately choose from meaningful symbolic systems ways of acting that reject it. I can decide not to have children, and live accordingly, and this, it is claimed, is evidence that we are not under the power of natural selection. They frequently also argue that culture itself is something that eludes natural selection. Cultures emerge, it is suggested, because of their truth, or their beauty, or just for some inexplicable essence of humanness, and individuals adhere to them as they will, at least in the modern world, leaving one culture for another, putting together in hodge-podge fashion their own unique culture by selectively sorting through this, that, and another existing culture to choose according to their own tastes.

It is undeniable that the ability to structurally alter the genome technologically is a potential game changer, and in fact we were already doing this long before the advent of CRISPR. Selective breeding is something humans have done with non-human animal populations as well as with our own species for centuries, although the latter variety has been limited by moral and political rules. However, to characterize these techniques, especially the cruder among them, as escape from natural selection is rather a stretch. The same equation of the effect of ecologies and the resources they present to organisms will be of paramount importance in determining how well humans with specifically altered genomes function. Choice about reproduction too must be understood in a more complicated way. The argumentative framework for natural selection takes it for granted that selection will operate in whatever way is most easily achievable, and so getting organisms that reproduce sexually motivated to do that will not have to do with giving them the cognitive abilities to figure out why their interests are really reproductive, however much they might be fooled by other impulses to believe otherwise. Indeed, even knowledge that sex leads, in some cases at least, to reproduction is not a necessary element of an organism with an effective behavioral apparatus for getting reproduction accomplished. Sex is pleasurable precisely because the consequence it naturally produces is so important, and pleasurable consequences of a behavior are a good way to motivate an organism to perform the behavior. The mere fact that in modern societies, where women are in the work place in roughly the same percentages as men and raising children is expensive and in some clear ways against the cultural *Zeitgeist* of hedonist individualism, we might find a significant number of men and women who consciously decide to avoid reproducing is not sufficient to show that culture has separated us from the core of sexual reproduction as primal selective force driving our behavior. For that, you would minimally need to show significant numbers of human beings who not only did not reproduce but were wholly unmoved by and uninterested in sexual behavior of any kind, that is, in the mechanism by which natural selection ensures that properly constructed humans pursue the behavior that will lead to reproduction. Is there any evidence that men and women are becoming less interested in sex? Perhaps. The numbers of the new identity category "asexual" seem to have grown somewhat in recent years. It is still however a tiny minority of the human population, and there is no obvious reason to believe it will get much larger than it now is. And how much of American and Western culture more broadly is concerned with sexuality? There are entire academic fields dedicated to the study of advertisements, popular culture, and related cultural phenomena that acknowledge the absolute saturation of such cultural material with this most basic fact of human life. There is no evidence at this point at least—however much the small percentage of the population that consciously decides not to have children grows—that this is likely to fundamentally change.

Culture Evolves

Culture must also be understood as something that is itself subject to the same evolutionary forces that drive the rest of human activity. It is quite commonly assumed in mainstream social science that it is culture that has catapulted humans out of the rules of natural selection. Nothing could be further from the truth. Indeed, culture can be demonstrated to evolve in accordance with forces in the natural and social environments. Culture must be understood at its very core as an adaptive mechanism. If culture were gratuitous to reproductive fitness, it would be a difficult matter indeed to try to explain why it could come about in all its spectacular complexity and ubiquity. When we see non-human animal culture, it is clear how centrally reproductive fitness is involved. Whale song is believed by nearly all who study it to have to do with mating. Chimp termite foraging tools and techniques are engaged with the procuration of food, i.e., somatic preservation, which is in evolutionary theory a concomitant of reproductive fitness. Human culture varies in degree here, not in kind.

Cultures exhibit all the characteristics necessary for natural selection. They vary, they are in competition with one another, and they can be passed down to descendants, though not apparently at the biological level of the gene. What the basic units of culture are is still being worked out. The field of memetics, inspired by Richard Dawkins' famous theorizing in *The Selfish Gene* (1976), offers some promise here, but the question is not settled. It is nonetheless clear that culture fits all the basic requirements to be understood as subject to evolution. Cultural change can be independent of some but not all the effects of natural selection. Those genetic variations that produce functional plasticity, that is, that produce different phenotypes depending on differences in environment, are involved in cultural change. A good deal of the confusion on the issue of the relationship of the genetic to the cultural has to do with the effort to talk about some human activities and traits as "purely cultural" and others as "purely genetic." As I have noted previously, the reality is that nothing human is either wholly cultural or wholly genetic. As Richard Alexander puts it, "Neither is true: the environment contributes to the expression of all behaviors, and culture is best described as a part of the environment" (1987:23–24). As an illustration, he looks at an example from the anthropologist Marvin Harris, who tries to explain a practice of female infanticide in elite castes in stratified societies by reference to the pursuit of economic and political power by the individuals involved and not by reference to reproductive success. Harris fails to recognize that he is stuck at the proximate level. Ultimate interests always involve reproductive success, and proximate results such as the happiness produced by status or the pleasure of sexual activity itself are simply the mechanism by which natural selection operates in animals like us.

Responding to the Culturalist's Critique: Marshall Sahlins vs. Sociobiology on Culture

A monumental figure in cultural anthropology, Marshall Sahlins, has provided the model response from the culturalist perspective to an evolutionary theoretical approach to the study of human culture. In his short book, *The Use and Abuse of Biology* (1976), Sahlins argues that a natural selection framework for studying culture "challenges the integrity of culture as a thing-in-itself, as a distinctive and symbolic human creation" (ibid.:x). This much is true. As I have argued, the evolutionary perspective on culture does insist that it be denied the pedestal it typically is awarded by those like Sahlins, as a human-specific activity independent of the most basic behavioral parameters that apply to all other life on the planet. The manner in which Sahlins goes about trying to defend his view of culture against the view of Darwinian evolutionary theory is instructive, and describing some of the features of his argument and their scientific weaknesses is as good a way as any of drawing our discussion of culture to a close.

Sahlins begins with a long passage in which he attacks the notion that war in humans can reasonably be seen as caused by the human propensity for aggression. His argument consists mostly of simply listing examples of possible cultural justifications for a war—honor, guilt, love, and humaneness are among those provided. But cultural justifications of war do not necessarily tell us anything about ultimate causation. These are efforts to provide a meaningful narrative about the war's necessity to members of the warring society. As we have already seen in the discussion of altruism in chapter four, culture has myriad ways to speak to us as to the reasons why we should behave "selflessly," even while it is patently clear when we analyze real situations that altruistic behavior almost always benefits the altruist, contrary to most of the cultural narratives about altruism. The same holds for war. Soldiers may be called to join the fray by appeals to a moral need to stop an evil aggressor, or to patriotic love of fellows and of country. The test would be to investigate the emotional and psychological state of soldiers in combat situations, and to compare the evidence gathered to similar baselines for individuals engaged in straightforwardly aggressive violence.

Sahlins also claims that sociobiological frameworks fail to reckon with the fact that wars do not involve individuals fighting with individuals for reproductive interests. They involve states competing with states, and, quoting Rousseau, he claims that the individuals involved in fighting a war cannot be pursuing reproductive interests because states mobilize them to fight based solely on their membership in the group. This demonstrates an almost complete failure to comprehend the evolutionary manner in which more primordial contexts of human aggression are linked to more contemporary ones. Humans lived for most of our time on the planet in small groups in which membership was defined by relatively close familial, that is, genetic relationship. Fighting to defend the group, then, or to advance the group's interest against the interests of another group was not determined fundamentally by an individual's choice to defend his individual interests. By definition, his interests are more tied to those of his group than to those of the group against which his goes to war. Groups that had considerable numbers of individuals emotionally wired in such a way as to resist the commands of leaders to go to war would not have done well in competition with others with individuals wired to act according to the more common behavioral predilection of more or less automatic ethnocentric sensibility. And what do wars accomplish, at bottom? They advance the interests of the victorious groups, whether in the acquisition of territory, or captive slaves, or economic dominance over the vanquished in the form of reparations payments and favorable trade relations. Once one recognizes, as I summarized in following Richard Alexander in chapter four, that group-on-group competition became the central component of threat to the realization of individual interests at some point in our distant past, we easily see that such competition is a matter of importance to the individuals in groups who are seeking to pursue their own reproductive interests.

The centerpiece of Sahlin's critical effort concerns kinship. He claims that kinship systems in primitive societies are not based on genetic degree of relationship, but on purely cultural factors such as residence or idiosyncrasies of the group in question. He runs through a litany of examples in pre-modern societies of adoption of the children of vanquished foes in war, fathers unblinkingly accepting parental responsibility for children conceived by their spouses in their absence that cannot be theirs, and communal living arrangements with significant promiscuity that would make it difficult to know the identity of fathers of children (1976:37–51). Such practices prove, in Sahlin's view, that kin selection cannot be operative in humans. Indeed, he thinks there are no human societies that adhere to the logic of kin selection:

> [T]here is not a single system of marriage, postmarital residence, family organization, interpersonal kinship, or common descent in human societies that does not set up a different calculus of relationship and social action than is indicated by the principles of kin selection.
>
> (ibid.:26).

Even if humans wanted to act according to the principles of kin selection, they would not be able to effectively do so. How would members of primitive societies without concepts of fractions in mathematics be able to calculate degree of relatedness of relatives? This claim demonstrates nothing more than an astounding ignorance of the most basic logic of the operation of natural selection. It is not required for the organisms in question—and this holds true for humans as well—to consciously pursue reproductive interests as primary, and indeed such a mechanism for producing adaptive behavior would be clunky and ultimately ineffective. Selection will operate in such a way as to advantage those who behave in greater alignment with reproductive interest and in doing so it will tend to change populations in ways that reflect the effect of selection. Given that selection was already operating at a time in evolutionary history before there existed organisms capable of consciously being aware of their own interests, it would have had plenty of opportunity to produce mechanisms to further this kind of behavior that operated at a level below the conscious mind. To imagine that selection in human kinship systems requires conscious calculation of the contribution behaviors make to fitness is to have failed to understand an essential element of evolutionary theory. Sahlins here even contradicts the logic of his own culturalist framework for understanding how behavior works, which posits culture as a force that drives individual action from somewhere below the level of individual rational calculation and in accordance with the logic of the cultural system. Natural selection departs from both the theory of action that is based on rational choice actors making cost–benefit analyses before acting and the theory that instead assigns agency to the system itself. Individuals and only individuals act, but they frequently act under the motivation of interests of which they are completely ignorant or only very impartially aware.

Sahlins simply cannot know, vociferously though he might claim such knowledge, how accurately or inaccurately real kin systems reflect genetic relatedness, since we do not have the genomic data on those populations to run the test. We do know though that cultural anthropological accounts of such systems are notoriously incomplete and frequently unreliable, as they are biased by the fact that anthropologists rely ultimately on what their respondents tell them about what is happening. If those respondents do not themselves know what is motivating their behavior, and if they do not have any information on the genetic relatedness of individuals in families, how can they accurately report on that to the anthropologists? The insinuation that Sahlins makes that there is a more or less infinitely complex array of kinship possibilities is also false. Robin Fox (1989) has shown that the set of rules that define kinship system options is low, probably between six and twelve, and the total number of logical combinations available from this limited array of rules is quite finite. And in actual practice only a small number of the possible combinations seem to be in any kind of frequent use in known cultural systems. The hypothesis of an evolutionary sociology would be that a genetic relatedness test in these societies would likely demonstrate much greater genetic relation of those in close kin relations than is suspected by cultural anthropologists.

A broader theoretical failure in this critique can be seen in its way of understanding the role kin selection plays in the overall structure of natural selection. Kin selection is not the only mechanism by which natural selection operates, and so even a coherent criticism of the role kin selection plays in tribal societies—bolstered by real data on how individual action in familial decision-making violates the principle of kin selection, which again we do not have—cannot alone indict or vindicate the theory of natural selection. We would need to know what else is going on in societies wherein familial practices that seem separated from kin selection are operative. What if, for example, we compare the inclusive fitness of men who adopt children not theirs to those who do not? Do the former have greater reproductive success according to the terms of indirect reciprocity? That is, do their reputations in the culture grow sufficiently, as a result of their adherence to established moral rules about adopting the children their wives have procreated with other men or taking in the children of vanquished opponents in war, to enable them to better realize their interests than others who ignore such cultural rules and therefore have their reputations damaged? The ways in which the various elements of an ecological system work themselves out are generally highly complex, and the kind of final statement about how well a population achieves reproductive fitness that Sahlins so readily makes here after a discussion of a few pages of some loose ethnographic data requires a much more thorough analysis, beginning with reliable genetic data on the level of relatedness of individuals within the group.

Evaluating the fitness of given familial practices has to do with the overall ability of the society to meet ecological challenges. All the populations Sahlins discusses are small Oceanian societies. Even if it proved to be the case that his ethnographic summaries of their familial practices were correct, and kin selection were more or less wholly ignored by these groups, and no indirect reciprocity or other compensating benefits to fitness were discovered, the existence of a few tiny samples such as these hardly disproves Darwinian theory, as Sahlins suggests it must (Sahlins 1976:51). Small populations of an exceptionally numerous species that fall into practices that are erroneous in terms of fitness are conceivable, just as demonstrably altruistic individuals can be made sense of in Richard Alexander's analysis of moral systems. They are errors in terms of the theory, and they are predicted to be unsustainable. How long it takes for that fact to reveal itself, however, is informed by many factors. The evidence of the competitive disadvantage the kinds of societies Sahlins describes face in comparison to the much larger societies around them, which observe kin selection fairly straightforwardly, is crystal clear. It is not too much to say that the sole reason such societies have survived this far into modernity has to do only with the emergence of modern ideologies by which members of modern societies attain moral status by defending the imperative of protecting small, primitive societies such as the Rangiroans or Trobrianders against the huge competitive disadvantage they face *vis-à-vis* modern societies. How much longer they can survive has mostly to do with that ideology of more dominant societies, not with anything that they are doing internally.

Marshall Sahlins shows no curiosity about any of these problems in his argument. Indeed, he apparently is altogether unaware of the weakness in his case that they present. Sahlins is one of the figures I most admired in my early career as a cultural sociologist, and he is obviously a serious thinker, but it must be said that he shows here that he is patently unfit to carry out the critique of the theory of natural selection that he undertakes. This is a matter, we might say, of his too rudimentary understanding of the relevant language. He is fluent in culturalist social science, but his facility with the language of natural selection is akin to the knowledge of French of most American study abroad students spending a semester in Paris. He is sufficiently aware of some basics as to be able to articulate a few coherent phrases, but his knowledge is far too rudimentary to permit serious and detailed examination of the nature of and evidence for the theory.

In the concluding pages of his critique of the sociobiological view of human culture, Sahlins sums up the culturalist claim for exceptionalism: Human culture, specifically insofar as it is based in language, is not defined by "the function of communication… [but by the] structure of signification [that is, it does not only] convey information [but it also] generate[s] meaning" (ibid.:61). He then gives as an example of how human cultural systems differ from non-human animal communication of information the way in which the symbolic meaning of color hierarchies operates in human societies. There is, as he acknowledges, ample evidence that color categories are essentially universal across cultures (see, for example, Lumsden and Wilson 1981:43–48), but he insists this is not the proper way to understand the relationship of culture to any biologically-shared cognitive behaviors such as the ability to categorize colors: "Colors in cultures are semiotic codes: they are used to signify the differences between life and death, noble and commoner, pure and impure" (Sahlins 1976:66). There is no doubt that some cultural variation exists in the ways in which colors are taken as symbols for other things, but there are also considerable elements of universalism involved in the working of such meanings. Considerable evidence indicates the near universalism of the six Hering primary colors—black, white, red, yellow, green, and blue. Beyond this, though, at the semiotic level, we find a range of facts shared in essentially all known human cultures: Black and white are commonly taken symbolically as representing opposing moral categories, red and yellow are understood as "warm" and green and blue as "cool," both with an associated range of emotive and moral terms and symbols, and red is often taken as signifying strong emotional content (Kay 1999; Kay and Maffi 1999). Sahlins, like many another culturalist analyst, is so fixated on the trees of differences, however comparatively trivial and unimportant they are in the grand scheme of understanding basic human nature and behavioral repertoires, that he fails to see the forest of remarkable uniformity and shared essence.

As I noted at the outset of this chapter, I was an insider to the game that is Sahlin's professional expertise for some time, and so I have a fair amount of insight into the main trepidations and defense mechanisms of culturalist readers who are suspicious of the biosociological and evolutionary approach. I have given up that culturalist game. What happened? Basically, the slow but inevitable grinding away of hard facts. One by one, all the axioms on which cultural sociology is built were weakened and collapsed under me. "Only humans have culture": Not so. "Human cultures are wholly unique and there are no cultural universals": Incorrect. "The origins of culture cannot be adequately accounted for by the theory of natural selection": Untrue. "Culture does not evolve in a manner at all consistent with the rules of natural selection": False. "Human cultures are infinitely complex sets of narratives and stories carried around in the heads of members that inform the behaviors of culture members": Wrong.

As the elements of the culturalist faith fall apart under the axe of ever-accumulating evidence, one can perhaps hope that more sociologists and culturalist social scientists will move in the same direction I have. But again the very evidence of our natures as constructed by the processes of natural selection gives us pause in this hope. We were an emotional species before we were a reasoning one, and that order remains in the structure of our minds. Our connection to the established ways of thinking of ourselves, firmly insulated as it is by our elevated, prideful self-regard, will not easily be undone.

References

Alexander, Richard. *The Biology of Moral Systems*. New York: Routledge, 1987.

Barrett, H. Clark and James Broesch. "Prepared Social Learning about Dangerous Animals in Children." *Evolution and Human Behavior*, 33 (2012):499–508.

Boyd, Brian. *On the Origin of Stories: Evolution, Cognition, and Fiction*. Cambridge: Belknap Press, 2010.

Brown, Donald. *Human Universals*. New York: McGraw-Hill, 1991.

Cronk, Lee. *That Complex Whole: Culture and the Evolution of Human Behavior*. Boulder, CO: Westview Press, 1999.

Damasio, Antonio. *The Strange Order of Things: Life, Feeling, and the Making of Cultures*. New York: Pantheon, 2018.

Dawkins, Richard. *The Selfish Gene*. Oxford: Oxford University Press, 1976.

Dutton, Denis. *The Art Instinct: Beauty, Pleasure, and Human Evolution*. New York: Bloomsbury Press, 2010.

Fox, Robin. *The Search for Society: Quest for a Biosocial Science and Morality*. New Brunswick: Rutgers University Press, 1989.

Kay, Paul. "Color." *Journal of Linguistic Anthropology*, 9, 1/2 (June 1999):32–35.

Kay, Paul and Luisa Maffi. "Color Appearance and the Emergence and Evolution of Basic Color Lexicons." *American Anthropologist*, 101, 4 (Dec. 1999):743–760.

Martin, John Levi. "Life's a Beach But You're an Ant, and Other Unwelcome News for the Sociology of Culture." *Poetics*, 38 (2010):228–243.

Mesoudi, Alex. *Cultural Evolution: How Darwinian Theory Can Explain Human Culture and Synthesize the Social Sciences*. Chicago: University of Chicago Press, 2011.

Riley, Alexander. *Impure Play: Sacredness, Transgression, and the Tragic in Popular Culture*. Lanham, MD: Lexington Books, 2010.

—. *Angel Patriots in the Sky: The Crash of United Flight 93 and the Myth of America*. New York, NY: New York University Press, 2015.

Sahlins, Marshall. *The Use and Abuse of Biology: An Anthropological Critique of Sociobiology*. Ann Arbor: University of Michigan Press, 1976.

Symons, Donald. *The Evolution of Human Sexuality*. New York: Oxford University Press, 1979.

Whitehead, Hal and Luke Rendell. *The Cultural Lives of Whales and Dolphins*. Chicago: University of Chicago Press, 2015.

Epilogue

The Evolutionary End of Sociology?

Why do we do sociology? Why do we want to know about ourselves and our societies? Whence our curiosity about our natures and the origins of our behaviors, and the reasons for the shapes of our social worlds? The first approach to an answer would simply declare our desire to have the most accurate information available in thinking about what is possible, how best to guide and direct our lives together to minimize strife and maximize well-being. But the desire to try to make the kind of societies we think we want or should have alone is insufficient to explain the emergence of sociology, the science of society. There is—or once was—something about the nature of the way of knowing that characterized emergent sociology that made it a particularly attractive and helpful tool for imagining what kind of creatures we are and how we should live together.

Ultimately, we want to know about the world we inhabit—which includes us and our societies—because it is our nature to pursue knowledge. And we want to know how to live, and how to live together, because it is our nature to want to live well rather than badly, to want abundant life rather than deprived, impoverished life. Science as a way of knowing emerged with that same dual focus: Learn what is the case because it is good to know, and then use that knowledge to try to make human life better because it is good to have more pleasant lives.

How well is sociology answering to that desire to understand in order to order our lives well? One reasonably wonders what of utility contemporary sociology can offer to this project. It is not just that it is too misguided about the topics discussed in these pages. Its estrangement from the scientific project is now sufficient, and its commitment to the moralizing advocacy project that Christian Smith (2014) has brilliantly documented and described as the sacred project of American sociology is so entrenched that it frequently actively opposes efforts to think scientifically about humankind.

I think often of what happened to the great anthropologist Napoleon Chagnon—that is, of what was done to him by the national association of anthropologists to which he belonged as a member of the profession. In the very year of his retirement from his position, this scientific giant of anthropology was denounced by a journalist who accused him, without evidence, of heinous crimes. The professional association of anthropologists—led by a team of those who adhered to a vision of their discipline based in activism and moralizing rather than science—that should have defended him instead joined in the denunciation and endeavored to utterly destroy his intellectual reputation. Fortunately, there were still a few saner heads in the ranks of their discipline, and eventually the mendacity of the entire campaign against him became clear and Chagnon was exonerated, though he had to spend much time and energy to defend himself from charges that ought never to have been taken seriously. But the whole sordid story should be explored, in all its awful details, by anyone contemplating a career in academic social science who is critical of the contemporary orthodoxies found in those disciplines. It does not say much for the future of a social science vigorously dedicated to knowing how the world is, whether that knowledge pleases us or not.

As is perhaps only too clear from its content, this book comes out of a profound disillusionment, one might even say a loss of faith. In the two decades since I received my doctorate in sociology, I have come to believe that my discipline is off the rails, and dangerously, perhaps fatally so. I did not believe that twenty-five years ago, when I was happily engaged in the process of digging around in Parisian archives to gather material for a dissertation on a small slice of the discipline's origins. Although I was worried then about the frivolous direction of much of sociology, and the disappearance of great sociologists who had influenced me and who were replaced by much lesser lights, I believed there was evidence to indicate the possibility of a renaissance, if only the message of the need to return to the sources and the original scientific impetus of the discipline were heeded by enough young scholars.

Today, based on the available evidence, I think it is more likely that sociology will collapse under the dead weight of its current concerns than that it will willingly reform itself according to the blueprint of this book and the other, better books that preceded it. Sociology is by some objective measures in evident decline at present. The last decade has witnessed a steady drop in ASA membership, from nearly 15,000 in 2007 to just over 11,000 in 2019 (a drop of around 25%). This is the lowest number of affiliates of the national professional organization for sociologists in the United States in nearly 40 years. Open, acute criticism and even outright disdain of the discipline is rampant, more than occasionally originating in sources who are themselves sociologists (see e.g., Smith 2014 and Cole 2001).

My book can be read, then, by sufficiently morbidly-minded readers, as an obituary to a dead discipline, or at least to one that is in the process of dying. It is also an act of mourning the loss of the expiring entity, which I have loved. But, even if this view is true, let us not mourn too much. It is a properly evolutionist way of thinking of the matter to recall that everything that exists eventually faces some catastrophe or insuperable problem, typically an ecology ill-suited for it, and it then passes into extinction and leaves the way for other organisms. Why not so with sociology in the form it currently takes? In this spirit, then, I can perhaps hope to have made some little contribution to the human science to come that will succeed it.

References

Cole, Stephen (ed.). *What's Wrong with Sociology?* New York: Routledge, 2001.
Smith, Christian. *The Sacred Project of American Sociology*. New York: Oxford University Press, 2014.

Index

For Product Safety Concerns and Information please contact our EU
representative GPSR@taylorandfrancis.com
Taylor & Francis Verlag GmbH, Kaufingerstraße 24, 80331 München, Germany

www.ingramcontent.com/pod-product-compliance
Ingram Content Group UK Ltd.
Pitfield, Milton Keynes, MK11 3LW, UK
UKHW022354190926
475702UK00015B/2291
9780367750978